P9-CCF-741

Chemical Protective Clothing

Volume 1

Edited by
James S. Johnson, Ph.D.
and
Kevin J. Anderson

American Industrial Hygiene Association, Akron, Ohio

ACKNOWLEDGMENT

The editors and chapter authors would like to acknowledge the initial funds from the American Industrial Hygiene Association as well as ongoing financial support from the Hazards Control Department of the Lawrence Livermore National Laboratory for the preparation of volumes 1 and 2 of *Chemical Protective Clothing*. Each author's organization also provided resources that permitted the completion of the work.

International Standard Book Number 0–932627–43–9 (Volume 1)
International Standard Book Number 0–932627–44–7 (Volume 2)

American Industrial Hygiene Association
P.O. Box 8390, 345 White Pond Drive
Akron, Ohio 44320

Contents

Contributors

James S. Johnson received a masters degree in Hygiene from the University of Pittsburgh, Graduate School of Public Health in 1972 and a Ph.D. in Chemistry from Duquesne University in 1972. He has worked at the University of California, Lawrence Livermore National Laboratory since graduation as an Industrial Hygienist, Safety Science Group Leader, and presently as the Acting Division Leader of the Special Projects Division of the Hazards Control Department. He has participated actively in the chemical protective clothing area since the late 1970s and he has also provided leadership to the American Society for Testing and Materials Committee F-23's standards process, which has generated the majority of current chemical protective clothing standards. He maintains an active research interest in all aspects of chemical protective clothing as well as many other areas of industrial hygiene. He is a certified Industrial Hygienist in Comprehensive Practice and a member of most professional and committees addressing chemical protective clothing.

Kevin J. Anderson received a B.S. with Honors in physics/astronomy from the University of Wisconsin, Madison, in 1983. He has worked for six years as a technical writer/editor for the Lawrence Livermore National Laboratory, assisting in the production of nearly all documents for the Hazards Control Department. He has also written and edited reports, brochures, and extensive documentation for the Air Force's Weapons Laboratory at Kirtland Air Force Base, New Mexico, and for three Technical Safety Appraisals performed by the U.S. Department of Energy. For two years he has written a monthly column for the Materials Research Society's *MRS Bulletin*. Mr. Anderson is also the author of nine novels published by Bantam Books and Signet Books.

Norman W. Henry III is a research chemist at the Haskell Laboratory for Toxicology and Industrial Medicine, Central Reseach and Development Department, E. I. du Pont de Nemours and Company, Inc. He received a Master of Science in Chemistry from the University of Delaware, 1977. He is a member of the American Chemical Society, the American Society for Testing and Materials, the American Industrial Hygiene Association, and the American Board of Industrial Hygiene. He is a Certified Industrial Hygienist (chemical aspects) and Clinical Chemist. He also has been certified by the National Certification Commission in Chemistry and Chemical Engineering. Mr. Henry has published several articles on chemical and protective clothing in analytical and industrial hygiene chemistry and he is currently active in protective clothing research. He is also the current chairman of the AIHA Protective Clothing and Equipment Committee.

Robert R. Jacobs, Ph.D. is an Associate Professor in the Department of Environmental Health of the School of Public Health at the University of Alabama at Birmingham. Included among his research interests is the development of models to assess the risk of systemic toxicity from dermal exposure. He received his Ph.D. from the School of Public Health at the University of North Carolina at Chapel Hill in 1972, and has had extensive experience in industry.

S. Z. (Zack) Mansdorf is President of S. Z. Mansdorf &Associates, Inc., a health and safety consulting firm headquartered in Cuyahoga Falls, OH. Zack received Masters degrees from the University of Michigan in Environmental Health and from Central Missouri State University in Industrial Safety. His Ph.D. is from the University of Kansas in Environmental Engineering. Zack is a Certified Industrial Hygienist and Certified Safety Professional. He has been a member of ASTM's Committee on Protective Clothing and the AIHA technical committee on protective equipment since 1977. He has published extensively on this subject and directed research in methods development for NIOSH. He is also a member of most professional societies addressing chemical protective clothing.

Jimmy L. Perkins, received his Ph.D. in Environmental Health Sciences from the University of Texas School of Public Health in 1981. He is now Associate Professor at the University of Alabama at Birmingham. He has performed research in the area of chemical protective clothing since 1983 and has numerous publications in the field, including peer-reviewed journal articles, book chapters, and edited books. He is a Certified Industrial Hygienist in Comprehensive Practice and a Certified Safety Professional.

Michael Roder received his B.S. in chemical engineering from West Virginia University in 1965. Since 1972 he has been a commissioned officer in the U.S. Public Health Service assigned to the National Institute for Occupational Safety and Health. As part of these duties he has been involved with chemical protective clothing research and evaluation and the development of a computer database summarizing test results since 1982. He is an active member of the ASTM Committee F-23 committee on protective clothing and the AIHA Protective Clothing and Equipment Committee.

Nelson Schlatter has a Bachelor's degree in Chemistry from the University of Delaware and 20 years of experience in industrial chemistry. For the last ten years, he has supervised the chemical resistance testing program at Edmont, a major industrial glove manufacturer, recently merged to form Ansell Edmont. He is a member of the American Industrial Hygiene Association Protective Clothing and Equipment committee, and of ASTM Committee F-23 on Protective Clothing. He has participated in the drafting and editing of ASTM F-739, the commonly accepted standard for permeation testing.

John Varos was the technical director for Mapa/Pioneer for years. In that capacity he oversaw the needs of a large multiproduct glove manufacturer. He was also a member of a large number of professional organizations related to the manufacture and use of chemical protective clothing, such as the Work Glove Manufacturers Association, the American Industrial Hygiene Association, and the Health Physics Society. He is currently the Ontario Operations manager for Smith Nephew Inc.

Chapter 1

Introduction

James S. Johnson, Norman W. Henry, III,
and Kevin J. Anderson

The use of protective clothing can be traced back to the beginning of civilization. Primitive people first used animal hides and furs to protect themselves from skin abrasions and cold; later, they learned to use a leather guard on the forearm as protection from bowstrings while hunting. Soon thereafter they learned to use wool from sheep to keep warm.

Early in the 18th century, people discovered the uses of cotton and rubber plants. Natives were known to waterproof articles of clothing and footwear with the gum from rubber plants cured over smoky fires. In the 20th century these natural sources of protective clothing were superseded by the discovery of synthetic materials such as neoprene and butyl rubber. Today, there are numerous synthetic plasticizers, elastomers, coated materials, and laminates to choose from.

All clothing is protective to some extent and for some length of time. Different types of clothing can protect from injury, cold, heat, water, pressure, or combinations of these elements. Perhaps the most complex type of protective clothing guards against exposure to harmful chemicals. No item of chemical protective clothing can provide complete protection indefinitely. To properly protect workers against the particular chemicals to which they are exposed, the health and safety professional must know the degree and duration of the protection offered by any resistant material. This book addresses the effects of hazardous chemicals on the

This farm worker is dressed in a manner that will allow safe pouring of concentrated pesticides. If an accident does happen, he can quickly wash the chemicals off his chemical protective clothing.

Some industrial jobs are extremely dirty. Chemical protective clothing can make these tasks safe to complete.

skin and body, and ways to prevent or minimize exposure by selecting or designing suitable protective clothing combinations made from adequately tested barrier materials.

The United States uses approximately 65,000 different chemicals in various situations. An additional 1000 chemicals are introduced into commerce each year. For approximately 70% of these chemicals minimal (if any) toxicity data is available. Therefore, a complete health and safety assessment is possible for about 2% of all these chemicals, and a partial assessment is possible for an additional 12%. Most of the toxicity information available is for drugs or pesticides. One can generally expect to find many industrial chemicals that have neither adequate toxicity data, nor chemical protective clothing (CPC) performance data.

Starting in the early 1970s much emphasis was placed on the carcinogenic or mutagenic nature of chemicals. Today we find a new concern—blood-borne diseases caused for the most part by contact with human fluids, such as blood or semen. The diseases of concern are HIV, AIDS, and Hepatitis B. In early October 1989 a task group of the ASTM F23 Protective Clothing Committee formed a subcommittee to develop performance test methods for any protective clothing to be used against body fluids.

Occupational safety and health professionals are beginning to recognize the implications of skin exposure to toxic chemicals. In 1985 the Bureau of Labor Statistics of the US Department of Labor found that out of 125,000 new cases of occupational illnesses among workers, more than 66% involved skin disease or disorders associated with repeated trauma (this has increased from 43% in 1980). The National Institute for Occupational Safety and Health

Asbestos abatement activities have significantly increased the use of fiber-resistant protective clothing.

(NIOSH) has included dermatologic disorders on its list of ten leading occupational health problems. Occupational skin disease prevention should be a priority item for the health and safety professional in the 1990s.

The Occupational Safety and Health Administration (OSHA) has recently addressed CPC in their Standard covering "Hazardous Waste Operations and Emergency Response," 29CFR Part 1910.120. This Standard requires a formal personal protective equipment program similar to the program required for respiratory protection.

Chemical protective clothing is available in many different forms and materials of construction. The two extremes which best illustrate this are the simple latex finger cot for finger protection, up to a Teflon-coated fiberglass totally-encapsulating chemical protective suit (TECP). The jobs for which each of these CPC items are designed are vastly different. One can cost 20 cents in bulk and the other can cost as much as $4500. Each item is very important in reducing a worker's individual exposure. Chemical protective gloves are one of the CPC items whose performance has been studied in great detail, which has provided a large database of permeation performance data. Other widely used CPC items are boots, coveralls, TECP suits, and respirators, all from various commercial manufacturers. This equipment forms the foundation on which the health and safety professional must establish a CPC program.

A pair of gloves, chosen correctly, can prevent harmful exposures to toxic chemicals.

It is often very difficult to select chemical protective clothing—frequently the health and safety professional has no permeation or breakthrough data to work with (the Appendix volume of this textbook should prove helpful in this regard), and many particular formulations of barrier materials are trade secrets, which makes it difficult even to *look* for permeation data. In other situations, the combinations of chemicals and particular exposure conditions of temperature, humidity, fibers, or particulates in the workplace, or the potential for physical damage to a garment are not readily adaptable to generalizations.

The health and safety professional must be ready to gather all the information available, carry out laboratory tests, if necessary, and use professional judgement in making the best CPC selection possible. Follow-up personnel monitoring and regular evaluation of CPC items are both necessary to confirm the performance of the selection. The health and safety professional must also observe to ensure that CPC items are being used properly by the worker. A recent situation demonstrates that proper selection, performance, and education do not always assure effective use: Researchers at Johns Hopkins University in Baltimore, MD, found that few emergency room personnel at the university hospital make use of available AIDS-protective gowns, gloves, and masks, despite a known HIV infection rate of 6–18% in emergency room patients.

This book provides a comprehensive treatment of the major aspects of chemical protective clothing design, performance, and use. It begins by addressing the effects of hazardous chemicals on the skin and the types of injury against which CPC can protect, then discussing how chemicals can permeate the skin. The next two chapters discuss basic polymer chemistry and how different protective materials are made, as well as a review of permeation theory, which explains how harmful chemicals can penetrate these barrier materials. Chapter 5 covers specific test methods that can be used to determine the actual protection potential of different barrier materials in the work environment. Chapter 6 describes the different types of CPC, giving details on the various methods of construction and manufacture of gloves, suits, boots, etc. The book then turns to actual CPC selection guidelines, using the data in the Appendix volume and Chapter 9 to help the health and safety professional plan, develop, and implement an effective chemical protective clothing

During the various Apollo missions, protective clothing provided astronauts with a portable and safe environment in space. The use of these systems on the Moon illustrates how comprehensive and effective protective clothing performance can be.

program. Chapter 8 covers methods for decontamination of protective clothing, allowing for a more efficient and cost-effective CPC program, when possible. The final chapter describes the key elements necessary to develop a chemical protective clothing program, which is essential to document the CPC selection process.

A recent article in the *Journal of the American Medical Association* demonstrates the increased attention devoted to CPC. The study involved the occurrence of non-Hodgkins Lymphoma (NHL) in farmers who had used a particular herbicide. They found that the greater the number of days the herbicide was used, the greater the risk. The study also reported an overall decrease in NHL for farmers who wore chemical protective equipment. This type of analysis of CPC use, with corresponding reductions in risk and disease, can be expected to become more common in the future.

If we examine how effective protective clothing can be, the NASA space suit program provides the best example of an absolute encapsulating environment under the most extreme conditions.

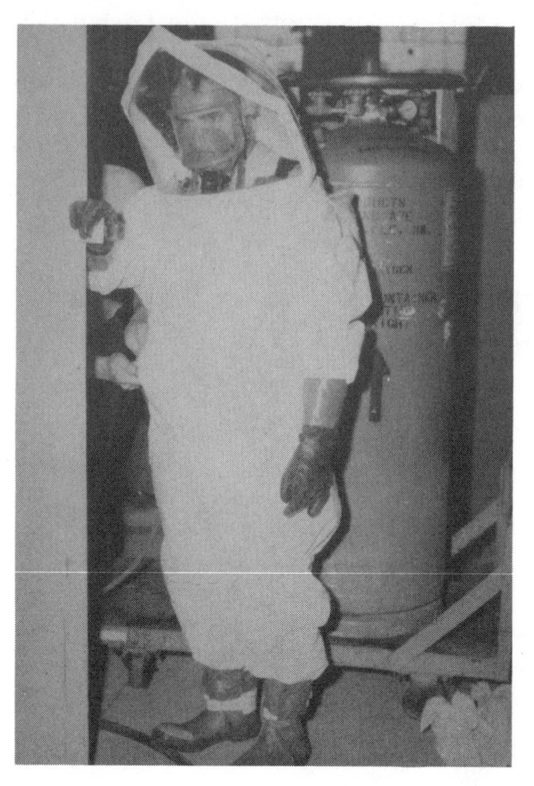

The US Army developed a totally-encapsulating chemical protective suit for application in their demilitarization cleanup of nerve-agent plants and facilities.

The US Army has also developed an effective nerve-agent demilitarization encapsulating suit which has been successfully used against some of the most deadly chemicals known.

What will the future hold? A recent development by the US Army illustrates a state-of-the-art technique on how CPC can be tested. A robotic mannequin "Manny" that not only looks like a human being, but can replicate human body functions such as sweating, has been developed to test protective clothing in conditions too hazardous to risk human subjects. The cost of this mannequin and related monitoring and control system, was $2.38 million.

This book will be useful to health and safety professionals, users of CPC, as well as chemists and manufacturers of chemical protective clothing. As industry and research expand into new areas, and workers must be protected against new hazards in the work environment, the need for comprehensive chemical protective clothing programs, in addition to test data and new barrier materials, will continue to grow for the foreseeable future.

A robotic mannequin named "Manny" developed by the US Army will permit the evaluation of chemical protective clothing in highly toxic environments. Significant improvements in CPC performance should result from these types of tests.

For Further Reading

S.M. Watkins, *Clothing: The Portable Environment*, Iowa State University Press, Ames, IA (1984).

R.L. Barker and G.C. Coleta, Eds., *Performance of Protective Clothing*, American Society for Testing and Materials, Philadelphia, PA, ASTM STP 900 (1986).

S.Z. Mansdorf, R. Sage, and N.P. Nielson, Eds., *Performance of Protective Clothing: Second Symposium*, Amercan Society for Testing and Materials, Philadelphia, PA, ASTM STP 989 (1988).

Chapter 2

Overview of Skin Structure, Function, and Toxicology

Robert R. Jacobs

Introduction

Occupational skin disorders are the major cause of injury and disability in the workplace. In the 1984 Bureau of Labor Statistics annual survey, 34% of all reported cases of chronic occupational diseases were dermatological disorders.[1] This is likely to be a 10- to 50-fold underestimate of the actual incidence, because the risks of disease from dermal exposures are poorly defined, and consequently association with occupation is difficult. Most of the above injury reports describe obvious skin disorders such as dermatitis, and only recently has more attention been given to skin permeation as a significant route of systemic exposure for toxic chemicals in the workplace and in the environment. Evidence from studies evaluating skin application of therapeutic drugs, as well as recent studies assessing skin penetration of solvent vapors in humans[2] and animals,[3] have demonstrated that the skin is potentially a significant route of exposure for systemic toxins. The need for a better understanding of the contribution from skin permeation to delivered dose is necessary not only to assess the increased risk from workplace exposure and to provide retrospective exposure assessment, but also to develop adequate chemical protective clothing.

This chapter will briefly review the structure and physiology of the skin, the pathology of skin disorders, and current concepts of skin permeation. Recent studies evaluating the risk of dermal exposure from workplace chemicals will be reviewed, and guidelines for assessing the potential for dermal permeation by industrial solvents will be summarized.

Structure of the Skin

The skin is the largest organ of the body, constituting 10–15% of the normal body weight, with a surface area of approximately 1.8 m^2.[4] Included in the functions of the skin are temperature and water regulation, as well as protection from the environment.

The skin is composed of two different layers, a thinner outer *epidermis* and the thicker underlying *dermis*. Also contained in the skin are several anatomical structures (appendages) that may be portals of entry for agents that penetrate the skin (percutaneous penetrants). These appendages include hair follicles, sebaceous glands, and sweat glands.

The Epidermis

The epidermis serves as the skin's primary barrier to non-traumatic disease. It consists of several distinct layers and ranges in thickness from 17–500 µm[5] with an average thickness of 40–50 µm (see Figure 2-1). The innermost layer is the *stratum germinatum*, or basal cell layer. This layer is metabolically active and is made up of a single layer of

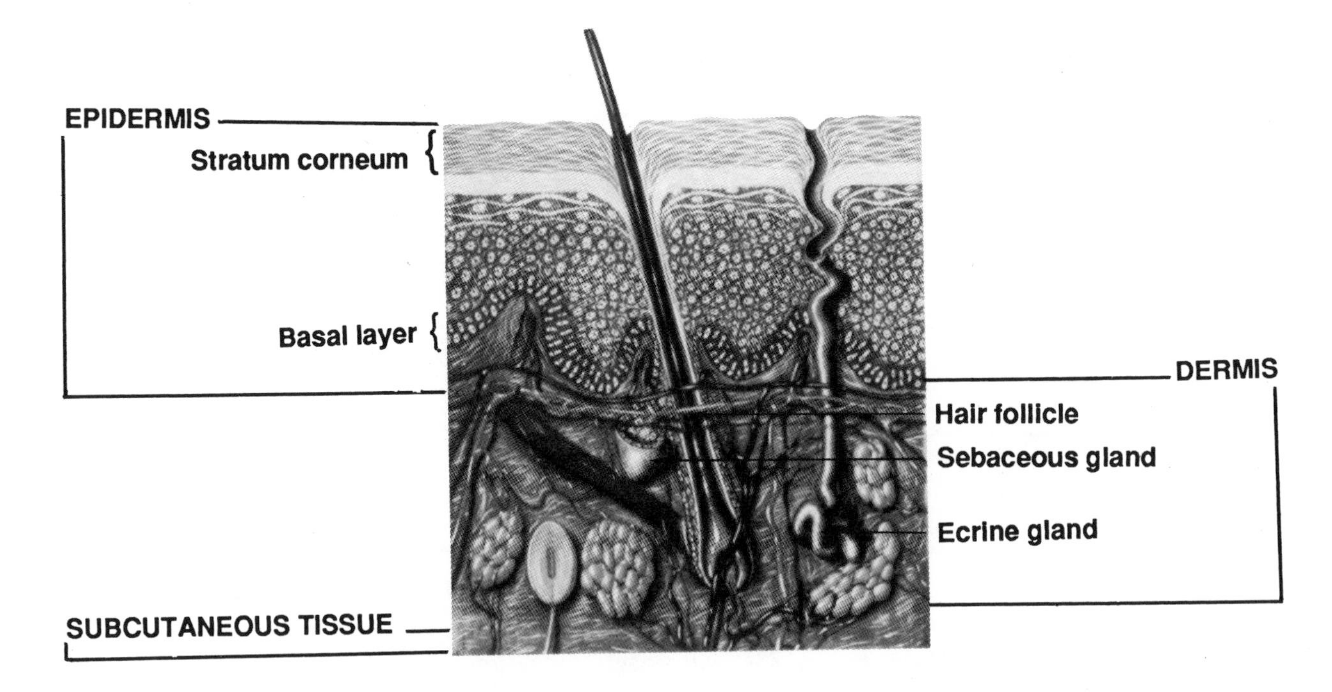

Figure 2-1. Major components of skin structure.

columnar epithelial cells, which through cell division is the source of all the cells of the epidermis. As cells of the basal layer divide and are pushed toward the surface, their physiological transformation from viable to dead cells forms the basis for each successive layer of the epithelium.

The next layer is the *stratum spinosum*, or the prickle cell layer. The stratum spinosum together with the basal cell layer are called the malpighian layer because of the presence of pigment-producing melanocytes. Melanocytes convert tyrosine into melanin, which provides some protection against the effects of UV radiation. As the cells of the stratum spinosum migrate toward the surface, they enlarge and produce cytoplasmic granules containing keratohyline, a precursor of keratin. These enlarged cells eventually die and make up the next layer called the *stratum granulosum*.

Further transformation of keratohyline to eleidin distinguishes the next layer, the *stratum lucidum*, from the stratum granulosum. The surface layer is the *stratum corneum*, which consists of several layers of corneocytes (dead epithelial cells) bound together by an intercellular lipid matrix. The time for transformation from the living basal cell layer to the dried, compact stratum corneum is 13–28 days.[6] As cells migrate from the basal cell layer to the stratum corneum, they lose water and become more compact, further enhancing the barrier properties of the skin.

The Dermis

The dermis consists of a fibrous network of blood vessels, collagen, and elastin divided into two layers—the papillary layer and the thicker reticular layer. Also contained in the dermis are several cell types involved in immunological, neural, and metabolic functions of the skin. The papillary layer interfaces with the basal cell layer through dermoepidermal ridges that provide an increased surface area for exchange of gases and nutrients and for temperature regulation (see Figure 2-1). The papillary layer ranges in thickness from 100–200 μm.[7] The thicker reticular layer of the dermis provides the structural stability and elasticity of the skin. Below the reticular layer is a layer of subcutaneous fat that provides padding and contains larger blood vessels and nerves.

Skin Appendages

Among the appendages found in the skin are hair follicles, sebaceous glands, and ecrine and apocrine sweat glands (Figure 2-1). Each is derived from the epidermal layer and extends into the dermis. The base of the hair follicle, the dermal papilla, is embedded in a tissue with a dense network of blood capillaries that is metabolically active. From this base an external root sheath, the *follicle*, extends to the surface of the skin. As with all the appendages, the follicle is lined with epidermis, thus the dermis does not come into direct contact with the environment. Cells of the dermal papilla divide rapidly, extending into the external root sheath. As these cells move toward the surface, they become keratinized, forming hair.

The sebaceous glands develop in the upper third of the hair follicle and extend into the dermis. These glands synthesize and secrete a lipid-rich material called *sebum* into the follicle, where it is transported to the skin surface through the external root sheath. This, along with other secretions, provides a thin lipid layer covering the stratum corneum, which provides limited protective properties for the skin.

Sweat glands consist of a secretory unit and an excretory duct lined with epithelial cells. The secretory unit is located below the dermis in the highly vascularized subcutaneous tissue and is connected to an excretory duct that follows a tortuous pathway through the dermis and epidermis to the skin surface. Ecrine sweat glands cover the entire body and secrete water, which both cools the surface and dissipates excess body heat. Apocrine sweat glands occur primarily in the auxiliary areas (groin and arm pits), genitalia, and nipples; these glands open into the external root sheath and secrete an odorless material into the hair follicle. The odor from apocrine secretions generally is a result of bacterial metabolites.

A more detailed description of the skin morphology may be found by consulting a textbook on dermatology.

Metabolic Capacity in the Skin

The primary barrier of the skin—the stratum corneum—consists of dead epidermal cells and is metabolically inert. For this reason the skin has historically been considered a passive membrane. However, the viable layers of the skin are active metabolically. This activity includes the basic cellular metabolism necessary for energy generation and maintenance of normal cellular function. In recent years researchers have recognized that skin contains metabolizing enzymes involved in the biotransformation of drugs or toxic compounds from the workplace. Biotransformation refers to the process whereby a chemical is changed to a more polar, more water-soluble substance that can be excreted more easily. Two main types of enzymatic reactions are involved in biotransformation: phase I functionalization reactions (oxidation, reduction, and hydrolysis) and phase II conjugation reactions. Table 2-1 lists some of the reactions that have been shown to occur in the skin.

Table 2-1. Biotransformation systems in skin.[a]

Oxidations	***Hydrolytic***
Alcohols	Epoxide hydrase
Aromatic ring	
Aliphatic hydroxylation	***Conjugations***
Alicyclic hydroxylation	Glucouronidation
Deamination	Sulfation
Dealkylation	Methylation
Reductions	
Carbonyl reductions	
C=C reductions	

[a]From T.A. Loomis, "Skin as a Portal of Entry for Systemic Effects" in *Current Concepts in Cutaneous Toxicity*, V.A. Drill and P. Lazor, Eds. (New York, Academic Press, 1980).

On a whole-organ basis, the skin contains about 2% of the total enzymatic activity of the liver.[8] This activity takes place in both the dermis and epidermis with the activity on a unit weight basis approximately equal for each tissue. However, because the dermis is the much larger layer, this is the site of most of the metabolic transformation that occurs in the skin.[9] This metabolic activity of the skin is inducible by a variety of compounds and is dependent on the type and quantity of inducing agent. While the total metabolic activity of the skin is less than that of the liver, it is nevertheless important when considering the permeation and ultimate toxicity of compounds applied to the skin. For example, while many topically applied compounds acted on by skin enzymes are detoxified, metabolism of the polyaromatic hydrocarbon benzo[a]pyrene results in the formation of a more-toxic species.[10]

Barrier Function of the Skin

The skin acts as a two-way barrier, preventing the entry of toxic agents from environmental exposure and preventing the loss of water, electrolytes, and other substances necessary to maintain homeostasis. The major barrier to permeation is the stratum corneum. This has been demonstrated in experiments showing that the skin permeation rate increased after each successive stripping of the stratum corneum with adhesive tape and in studies demonstrating the accumulation of permeants in the stratum corneum.[11]

Early studies of electron micrographs portrayed the stratum corneum as a highly porous tissue of layered plates of epidermal cells with no intercellular substance. It was concluded from this that the stratum corneum did not provide a barrier function. With improved methods for evaluating the ultrastructure of the stratum corneum, studies have now demonstrated (as shown in Figure 2-2) that the stratum corneum consists of overlapping layers of corneocytes with a continuous intercellular lipophilic matrix.[7] Each corneocyte consists of keratin filaments in a matrix of lipid and non-fibrous protein. This filament matrix appears to be important for the barrier properties of the stratum corneum. Each cell is surrounded by a tough membrane (15 nm thick) that is resistant to agents which destroy keratin. The intercellular spaces are filled by lipids released during keratinization and represent about 1% by volume of the stratum corneum.

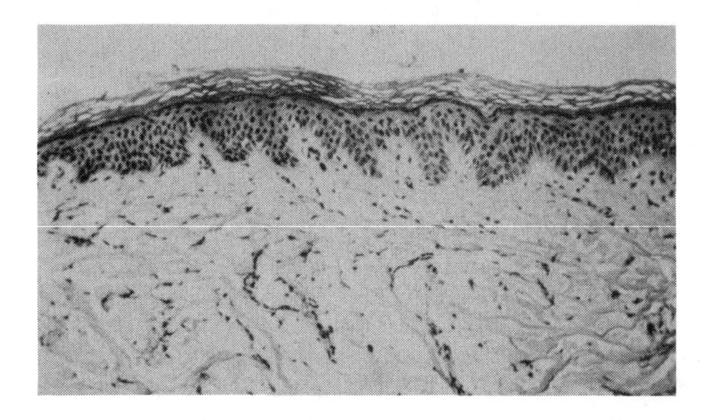

Figure 2-2. Electron micrograph of the epidermis, illustrating the ultrastructure of the stratum corneum. (Photo courtesy of Dr. Mitchell Sams, Chairman of Dermatology, University of Alabama, Birmingham, School of Medicine.)

Until recently it was thought that polar molecules permeated through the intracellular lipid.[7] However, studies now suggest that the intercellular route predominates for both polar and non-polar molecules: the transport of polar molecules occurs through the hydrophilic regions and the non-polar molecules occurs through the hydrophobic regions.[12] Regardless of the route of permeation, non-polar molecules permeate more rapidly than polar molecules, and molecules soluble in both polar and non-polar solvents permeate most rapidly. Exceptions to this order of permeation may occur if transport through hair follicles or sweat ducts predominates. This latter process is referred to as *appendageal transport*.

While the barrier properties of the stratum corneum provide the most substantial protection against agents that would permeate through the skin, other properties of the skin enhance resistance to microbial agents, physical trauma, thermal stress, UV radiation, and chemical trauma. Through the secretion and accumulation of sweat and dermal metabolites, the skin is covered with a thin lipid-based layer. The pH of this layer normally ranges from 4.5–6.[13] While offering no protection as a barrier against permeation or trauma, this residue will provide limited buffering capacity against weak acid and alkaline solutions. This layer also contains free fatty acids from sebum, which have demonstrated antimicrobial activity;[14] this coupled with the barrier function of intact skin prevents the invasion of microorganisms. Profuse sweating may provide a dilution effect for water-soluble toxins at the skin surface, but associated increases in hydration of the stratum corneum will likely overwhelm any dilution effect by enhancing penetration.

The skin also provides limited protection against UV radiation through the production of melanin. Exposure to UV radiation results in increased production of melanin that is dispersed to corneocytes in the surface layers of the epithelium, thus darkening the skin.[15] Prolonged exposure to UV radiation also increases the thickness of the stratum corneum, which may provide additional protection.

The collagen and elastic fibrous tissue of the dermis provides not only structural support for the skin, but also resistance to impact injury. This, coupled with the neural component for recognition of noxious stimuli and the immune surveillance system, provides the skin with a well-rounded defense system for responding to environmental stimuli.

Alteration of the Barrier

A number of conditions can alter the barrier properties of the stratum corneum and increase the rate of skin penetration. These can be categorized into factors that alter the permeability of the stratum corneum without damaging the membrane or specific factors that result in altered membrane integrity (see Table 2-2).

Hydration of the stratum corneum results in both increased thickness and larger membrane diffusion constants.[7] While these factors oppose one another, for most substances the larger diffusion constant predominates, and permeation is increased. Occlusion, a typical result of wearing chemical protective clothing, will both increase the temperature and hydrate the membrane, which will generally increase permeation.

Variations in thickness and appendageal density also influence membrane permeation. Regional variations in thickness of the stratum corneum range from 400–500 μm in the palms and soles to 20 μm or less for other areas. Increased thickness does not relate directly to lower rates of permeability.

Table 2-2. Factors that influence the barrier properties of skin.

Non-Damaging	*Damaging*
Hydration	Chemical Exposure
Appendageal Transport	Physical Trauma
Thickness	Disease
Temperature	
Exposure	
—Duration	
—Concentration	
—Area	

Scheuplein and Blank[7] ordered permeation rates for small molecules from highest to lowest as soles > palms > back of hand > scrotal/postauricular > scalp > arms, legs, and trunk. The role of appendageal transport in skin permeation has been evaluated both experimentally and mathematically. Tregear found no evidence of rapid entry of tri-n-butyl phosphate through hair follicles of pig skin;[16] however, other studies comparing permeation through skin sites with different densities of follicles found higher permeation rates in follicle-rich areas.[17] Scheuplein and Blank concluded that for most penetrants appendageal absorption is significant only in the early stages of permeation. This is based on the fact that the appendageal surface area is small (0.1–1% total surface) and the subsequent conclusion that absorption over the total surface eventually overwhelms appendageal permeation.

The length of exposure and the surface area exposed both affect the quantity of material that permeates the stratum corneum, as does the type and concentration of penetrant. Temperature also influences permeation by altering the membrane diffusion constant.[18] As temperature increases, permeation rates increase—this may be related to increased fluidity of lipids in the stratum corneum.

If the stratum corneum is altered by disease, such as psoriasis or eczema, it is more permeable than healthy skin.[19] The degree of altered permeation is dependent on the type and extent of disease and conditions of exposure. Damage of the stratum corneum from physical agents such as cuts or burns (including UV radiation) or chemical damage from solvents or detergents also alters membrane integrity and results in enhanced penetration of all compounds. On continued exposure to a damaging agent, the rate of permeation will increase to a point where only limited barrier function exists.

The contribution of skin metabolism to a local or systemic toxic response from dermally applied materials is so far established for only a few substances. As with toxic responses for other routes of exposure, factors that influence access to receptor systems control toxic response. For a toxic response to occur, the rate of transport across the skin to the receptor must exceed the rates of inactivation and excretion of the agent; the level of agent at the receptor site must also exceed the threshold of toxicity. Therefore, the barrier property of the skin is a major factor influencing the toxicity of dermally applied materials.

Occupationally Induced Skin Disorders

Occupational skin diseases have been defined as any skin abnormality induced or aggravated by the work environment.[4] This definition is directed toward those agents that damage the skin, including corrosives, irritants, allergens, and carcinogens. While a threshold of exposure must be exceeded for the following effects to occur, not all individuals are equally susceptible. Susceptibility is dependent on the person, the status of the stratum corneum, and the type and duration of exposure. This definition does not consider the risk of systemic toxicity from percutaneous absorption of toxic agents. Table 2-3 lists several skin disorders that may be caused by exposure to toxic agents in the workplace.

Table 2-3. Types of occupational skin disorders.

Contact Dermatitis	*Pilosebaceous Disorders*
Irritant	Acne
Allergic	Alopecia
Photodermatitis	*Pigment Disorders*
Photoirritant	Hyperpigmentation
Photoallergic	Hypopigmentation
	Leukoderma (Depigmentation)
Biological Dermatosis	*Cancers*
Viral	Basal Cell
Fungal	Squameous Cell
Parasitic	Melanoma

Irritant Dermatitis

Irritant dermatitis is a non-immunological localized inflammatory reaction following single or repeated exposure to a chemical.[20] Three types of reactions have been described. The first type results from a single exposure to a chemical, causing erythema and edema. Erythema refers to skin redness caused by damage to blood capillaries, which results in the influx of blood to the affected area. Edema refers to increased fluid in the tissue around the affected area.

The second type of reaction shows no visible sign on a single exposure, but repeated exposure results in a reversible erythema, chapping (fissures in the horny layer), erythema crackles (fissures in the stratum spinosum), and hemorrhagic fissures on the skin (cracks in the dermis). See Figure 2-3. This is the most common occupational contact dermatitis.

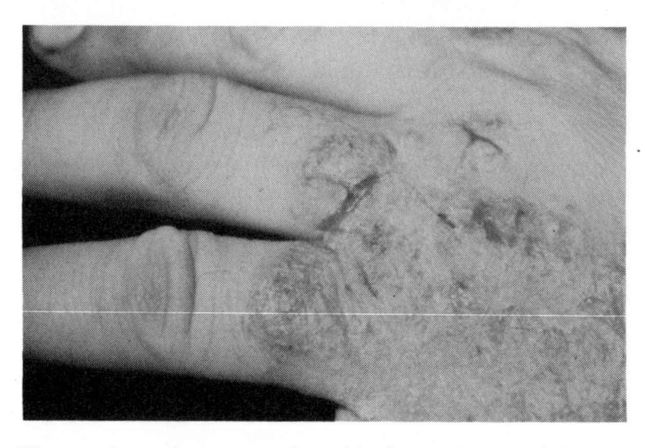

Figure 2-3. An example of irritant dermatitis.

The third type of response occurs on repeated exposure but results in a chronic dermatitis not easily reversible. At this time it is not possible to predict the intensity of reaction to one irritant by knowing the intensity of response to other irritants. The mechanism of irritant dermatitis is not known, but direct influence on vessel walls in the dermis or release of inflammatory mediators such as histamine may cause erythema and edema. Commonly identified irritants include detergents, various organic solvents, and acids and bases.

Allergic Contact Dermatitis (ACD)

Clinically, allergic contact dermatitis (ACD) is similar to irritant dermatitis, but the mechanism of the disease is from allergic sensitization. Figure 2-4 illustrates the steps required for ACD. Skin sensitizing agents are small-molecular-weight compounds (less than 500 MW) called haptans that combine with skin proteins to form an antigen that interacts with the T-lymphocytes of the skin's immune system.[21] On re-exposure, a similar antigen is formed and results in an amplified inflammatory response. The antigen penetrates the epidermis and interacts with antibodies on cell surfaces (e.g., mast cells). This interaction results in the release of histamine and accounts for the erythema and edema. Other cell mediators may also be released and influence the response. ACD may spread beyond the site of contact to a generalized response. The most common skin-sensitizing agent is pentadecylcatechol, the active agent in poison ivy. However, a number of occupational skin sensitizers have been identified, including chromium, cobalt, and nickel.

Phototoxicity/Allergy

Ultraviolet radiation is an important factor in inducing skin damage and disease. Ultraviolet wavelengths of 200–400 nm cause the most damage to skin because they are poorly screened by the atmosphere. Radiation in this range is divided into three regions: UV-A (315–400 nm), UV-B (280–315 nm), and UV-C (200–280 nm), each with a different spectrum of bioactivity. Biological effects of UV radiation include erythema, alterations in the dermal elastic tissue that results in actinic elastosis or aging of the skin, alterations in cell constituents such as DNA, which can cause cancer, and increased melanin production that darkens the skin. Apart from direct UV damage, dermal phototoxicity or photoallergy may result from exposure to chemicals that are activated by UV or visible radiation. These reactions are analogous to irritant and allergic contact dermatitis and differ only in the requirement of UV or visible light for activation. Phototoxic agents include several polycyclic aromatic hydrocarbons as well as selected drugs and may act through direct contact with the skin (e.g., anthracene) or after ingestion or injection (e.g., tetracycline). Those with occupational exposure to polycyclic aromatic hydrocarbons, coal-tar derivatives, or fungal byproducts may be at risk. Photoallergic reactions are rarer than phototoxic reactions. For photoallergic reactions the actual allergen is formed only by interaction with UV or visible radiation.

INITIAL EXPOSURE

Chemical (haptan) + Tissue protein = Antigen

Antigen + T-lymphocyte = Sensitized T-cell

RE-EXPOSURE

Sensitized T-cell + Antigen + Skin cell = Allergic response

Figure 2-4. Steps for chemical sensitization.

Follicular Diseases (Pilosebaceous Disorders)

Several disorders of the hair follicle can result from exposure to industrial chemicals. Abnormal hair loss (alopecia) has been reported on exposure to thallium or boric acid.[22] Far more common is environmental acne resulting from contact with petroleum derivatives, coal-tar products, halogenated aromatic hydrocarbons, and other chemical and biological substances. Chemical stimulation causes an increased proliferation of follicular epithelium that becomes keratinized, eventually plugging the hair follicle, which then retains sebum. Inflammation around the follicle—induced by the chemical itself, bacterial metabolites or infection, hormonal secretions, or immune responses—result in tissue breakdown and abcess formation.

The most notable environmental acne is chloracne caused by exposure to halogenated aromatic hydrocarbons (e.g., polychlorobiphenyls or dioxin). Lesions are typically straw-colored cysts accompanied by inflammatory pustules and abcesses and may occur over the entire body (both the exposed and non-exposed areas). The most common sites are around the eye and ear, with more limited involvement of the cheeks, forehead, and neck (see Figure 2-5). This response is generally related to systemic toxicity. Therefore, all routes of exposure appear to be important.[23]

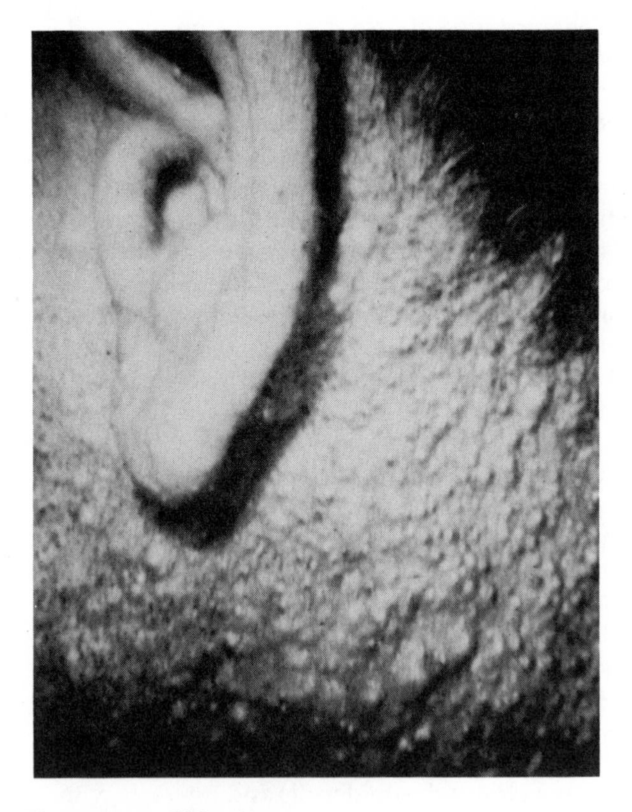

Figure 2-5. Chloracne.

Cancer

Cancers of the skin are divided into two groups: the non-melanoma cancers, which include basal cell and squameous cell carcinomas, and the melanomas.[24] Basal cell carcinomas are the most common skin cancers and occur primarily in persons with prolonged or intense exposure to sunlight. These growths appear as elevated lesions firm to touch on the nose, eyelids, cheek, and trunk, and rarely metastasize, or spread to other tissues. Less common are squameous cell carcinomas, which occur primarily on areas exposed to ionizing radiation, carcinogenic chemicals, or trauma. Their appearance varies from an ulcerated infiltrating mass (tumor) to an elevated, erythematous, nodular tumor. Squameous cell carcinomas may metastasize to regional lymph nodes. The least common but most difficult to treat skin cancers are melanomas. Melanomas arise from melanocytes and occur as pigmented blemishes with an irregular notched outline that can occur over the entire body (Figure 2-6). The incidence of melanoma is increasing significantly and it is often fatal, since it quickly metastasizes through the lymphatics.

Exposure to solar UV-B radiation is the primary carcinogen stimulus to skin, although exposure to chemicals either directly (e.g., polycyclic hydrocarbons) or through photoactivation (e.g., dimethylbenzanthracene) increases the risk of cancer.

Pigment Responses

Skin pigmentation can be altered by both physical (radiation) and chemical exposures. Hyperpigmentation is the most common work-related disorder and is caused by thermal or chemical burns or agents of contact dermatitis. Those working in the sun or with coal-tar products or with fruits or vegetables contaminated with fungi producing psoralens are at risk.[25]

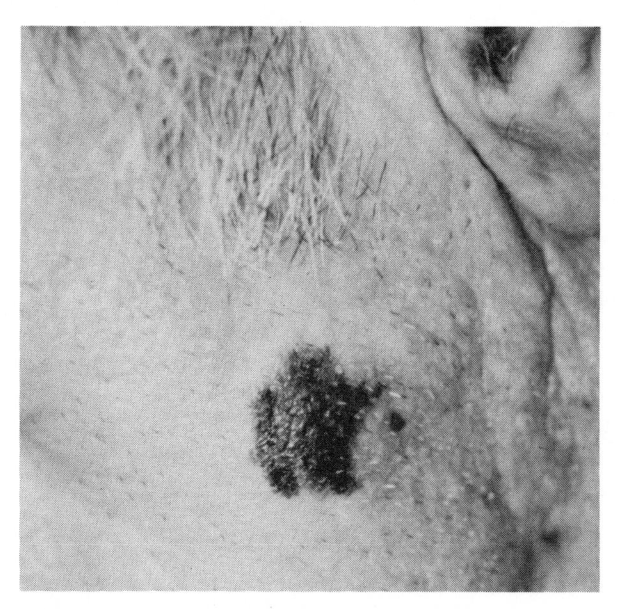

Figure 2-6. Melanoma. (Photo courtesy Dr. Mitchell Sams, Chairman of Dermatology, University of Alabama, Birmingham, School of Medicine.)

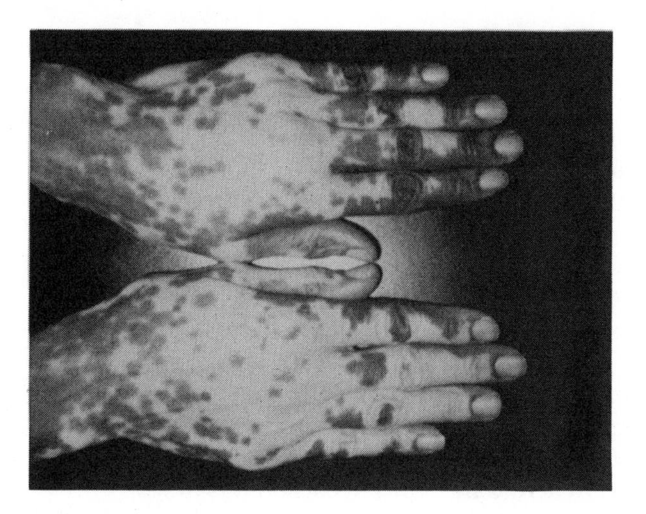

Figure 2-7. Depigmentation.

Decreased pigmentation also occurs on exposure to environmental chemicals. This is generally associated with more toxic or irritant effects that can destroy or affect melanocytes in the malpighian layer of the epidermis. Second- or third-degree burns and hydrofluoric acid are agents that can damage the skin and cause decreased pigmentation. A number of chemical depigmenting agents have been identified. Exposure to these agents induces leukoderma or pigment loss, as shown in Figure 2-7. First reported in workers wearing rubber gloves containing monobenzylether of hydroquinone,[26] a variety of alkyl phenols have also been associated with leukoderma.[27] The mechanism of depigmentation is not known. Structural similarities of the alkyl phenols to tyrosine, a building block of melanin, may interfere with enzymatic reactions that incorporate tyrosine into melanin.

Biological Skin Disorders

A variety of biological agents cause occupational dermatoses. These include agents of infection, infestation, and biological products that cause irritation or sensitization. While they are too extensive to review, the primary occupations at risk include agricultural workers and those processing animal or vegetable products.

Systemic Toxicity

As previously mentioned, the risk for systemic toxicity from skin exposure is not well defined. Although certain agents may damage the skin on exposure, others may permeate without apparent cutaneous damage. The remainder of this chapter will discuss the permeation of skin by organic solvents.

Skin Permeation

Permeation Theory and Assumptions

Scheuplein and Blank published an extensive review of skin permeability, which describes the mathematical basis of the permeation of dilute solutions through the skin, based on passive diffusion theory.[7] The application of passive diffusion theory to dermal permeation is based on the premise that the rate-limiting diffusion barrier is the stratum corneum. For this theory to apply, certain assumptions are made:

1. The stratum corneum is a uniform barrier.
2. The rate-limiting step of molecular movement is passive diffusion across the stratum corneum.
3. Clearance of the penetrant from below the stratum corneum is sufficiently rapid and does not affect the rate of absorption.
4. The stratum corneum behaves as a solution phase into which the penetrant must dissolve.
5. The structure of the stratum corneum is not altered by the penetrant over time or by changing concentrations of the permeant.
6. The driving force for movement of the permeant across the stratum corneum is the concentration difference.

A discussion of passive diffusion theory is given in Chapter 4. The following discussion describes the application of passive diffusion theory to the skin.

For dilute aqueous solutions Fick's law generally applies to the permeation of chemicals through skin. However, as the solute concentration approaches saturation and molecular interactions cause departures from ideal situations, Fickian diffusion does not apply. Several mathematical models have been developed to predict the flux of different solvents from saturated aqueous solutions through skin.[28-30] These models have used a number of parameters as the basis for predicting permeation including water solubility, molecular weight, molecular volume, vapor pressure, and different partition coefficients (e.g., the octanol:water partition coefficient). Although each has achieved limited success, these models cannot yet be used to predict permeation for broad ranges of chemicals. This may result from the stratum corneum not conforming to the assumptions described earlier or because the models do not incorporate all the variables that influence permeation.

As more is learned about the structure and composition of the stratum corneum, improvements in the mathematical models will be made. For chemicals that alter the barrier properties of the stratum corneum, neither Fickian diffusion nor the mathematical models will be useful for describing the permeation of concentrated solutions through skin. The following observations generally hold for diffusion of dilute solutions at steady state through the stratum corneum.

Permeability of Aqueous Solutions

Simple polar non-electrolytes penetrate the skin at rates similar to water. As the molecule becomes more polar by the addition of functional groups, the permeability decreases. This is a reflection of the smaller diffusion coefficient. The introduction of groups capable of hydrogen bonding may reduce the permeability constant to the point where appendageal transport becomes the predominant route for permeation. Introduction of non-polar groups (e.g., methylene) increases permeability, which is likely a result of increased solubility. Beyond a linear carbon chain of ten C, molecular size reduces mobility regardless of characteristics of the molecule.

Electrolytes do not penetrate the skin readily. This is likely a result of hydration of the ions forming a large diffusion unit and interaction of charged ions with charged functional groups in the tissue. Appendageal diffusion may be significant for electrolytes.

Dilute solutions of anionic and cationic surfactants penetrate skin readily. This may be more a result of damage to the stratum corneum than increased permeability constants.

Permeability of Non-Aqueous Solutions

Many non-aqueous solutions do not alter the skin membrane integrity, and diffusion theory applies to their permeation. However, low-molecular-weight volatile solvents, such as acetone, damage the membrane by dehydration or extraction of lipids, thus altering the diffusivity of the membrane and consequently cannot be described by passive diffusion theory. For organic solvents that do not damage the stratum corneum, lipid solubility appears to be the most important parameter in determining permeability. For this reason, oil:water partition coefficients have been used as the basis to predict permeability in many of the models of skin permeation. For example, the model of skin permeation proposed by Michael et al.[28] demonstrated a linear relationship between steady-state flux of selected drugs and the mineral-oil:water partition coefficient. Compounds with a higher mineral-oil:water partition coefficient permeated most rapidly. The octanol:water partition coefficient has also been used to estimate permeation, and this coupled with vapor pressure may provide a means to estimate the risk for skin permeation. For example, a compound with a low vapor pressure and a high oil:water partition coefficient may represent a risk for dermal permeation because it will not readily evaporate from the surface and is soluble in the liquid matrix of the skin. Examples of organic compounds that do not damage the stratum corneum include higher-molecular-weight alcohols and esters, as well as a variety of neutral high-molecular-weight compounds.

Certain organic agents are being used to enhance penetration of drugs without damaging the skin. Among the compounds being considered as penetration enhancers is the solvent dimethyl sulfoxide (DMSO). DMSO is miscible with organic and inorganic solvents. Other penetration enhancers include: 2 pyrolidon, oleic acid, mannitol, and hydrocortisone. Current thought suggests that these enhancers act by reducing diffusional resistance and promoting partitioning by reversible interaction with hydrophobic and hydrophilic regions of the stratum corneum.[31]

Permeability of Vapors

Little information is available to describe the permeation of organic vapors through the skin. For vapors of alcohols and alkanes, the diffusion coefficient decreases with increasing molecular weight. Addition of hydroxyl groups, while increasing the solubility of the alcohols, also decreases the diffusion coefficient. Diffusion constants for reactive condensable vapors are considerably lower than for the nonreactive permanent gases.[7]

Measurement of Permeation

Both *in vitro* and *in vivo* assays have been developed to measure percutaneous absorption of different chemicals and pharmaceutical agents. The most widely used *in vitro* assay is the diffusion cell (Figure 2-8). This cell consists of two parts, a donor chamber and a receptor chamber. Excised skin is placed as a barrier between the two chambers. The receptor cell contains a fluid over which the skin is placed, dermal side down, and the epidermal surface is dosed with the agent of interest. The fluid in the receptor can be exchanged at a specific rate and fractions can be collected at selected time intervals to obtain plots of percutaneous penetration vs time. From these data a permeability constant can be determined.

Whole-thickness and split-thickness skin, the epidermis as well as the stratum corneum from both human and animal skin, have been used as specimens for *in vitro* assays. Reasonable agreement with *in vivo* human studies has been obtained with excised human skin, pig skin, and skin from monkeys.[32] Skin from other animals, such as rats or guinea pigs, has generally shown poor correlation with the same *in vivo* human studies.[33]

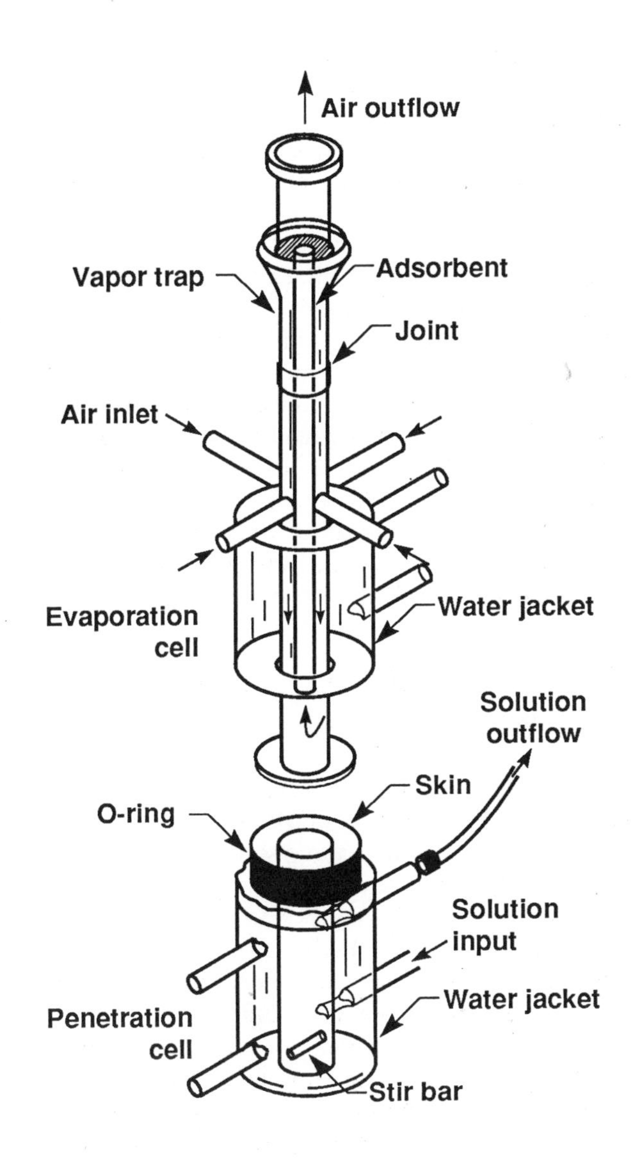

Figure 2-8. Illustration of an *in vitro* permeation chamber. (From Reference 33.)

In vitro assays have several limitations: The assay does not provide information about the effects of metabolism on the permeant. The use of a receptor fluid that differs substantially from blood may incorrectly model the partitioning of the permeant in the skin and blood. Exposure to selected agents may damage or hydrate the membranes, thereby altering normal permeation rates. Finally, shunt diffusion may be more pronounced in excised skin. However, the use of carefully controlled *in vitro* assays has provided a simple rapid method to measure diffusion kinetics and has produced reliable information, provided the limitations of the assay are acknowledged. Guidelines for *in vitro* skin permeation studies have been recently published.[34]

In vivo assays for percutaneous absorption have generally been done by dosing a defined area (e.g., forearm) with either a radiolabeled compound or an unlabeled compound that can be measured directly or as a metabolite in excreta, plasma, or exhaled air. The excretion of the agent is followed over time and compared with excretion after administration of the same compound by injection. The total skin permeation is the total dose excreted as a percent of the applied dose, corrected for the portion of the compound not excreted as determined by the injected dose. Feldmann and Maibach have published several papers describing *in vivo* percutaneous absorption of organic compounds in man.[35–37] These papers have been the basis of comparison for many *in vitro* assays as well as the basis for developing animal models of percutaneous absorption.

Percutaneous Absorption of Organic Solvents

***In Vivo* Human Exposures.** Stewart and Dodd published one of the first studies evaluating the permeation of skin by organic solvents.[38] Using human volunteers, they exposed skin on the hands by thumb immersion to carbon tetrachloride, trichloroethylene, tetrachloroethylene, methylene chloride, and 1,1,1-trichloroethane and measured the amount of the solvent in alveolar air. Comparison of these results was made to alveolar air levels after vapor exposure. Measurable quantities of each of the five solvents were detected within 30 minutes. Penetration was influenced by skin location, the surface exposed, and the duration of exposure. A quantitative comparison of vapor exposure and skin exposure suggested that only carbon tetrachloride was likely to permeate in toxic quantities and that toxic levels would be achieved only when the skin was submerged, as was done in this study.

Dutkiewicz and Tyras evaluated the permeation of pure liquid and aqueous dilutions of ethyl benzene in man by whole-hand immersion and topical application.[39] The rate of absorption of pure ethyl benzene was 22–33 $mg/cm^2/hr$; for aqueous solutions containing 112 and 156 mg/l, the rate was found to be 0.118 and 0.215 $mg/cm^2/hr$, respectively. The rate for absorption of pure ethyl benzene was much greater than that cited for aniline, benzene, nitrobenzene, carbon disulfide, and styrene. Dutkiewicz and Tyras concluded that exposure to pure ethyl benzene for one minute by whole-hand immersion was equivalent to an 8-hr exposure to airborne levels of 23 ppm. In a follow-up study, the absorption rates of toluene, styrene, and xylene topically applied to the forearm of human volunteers were 14–23, 9–15, and 4.5–9.6 $mg/cm^2/hr$, respectively.[40]

Feldman and Maibach reported on the percutaneous penetration of 21 different organic chemicals.[37] Radiolabeled chemicals were applied to the forearms of human volunteers, and the permeation was determined by measuring the appearance of the label in urine. Differences in maximum absorption rates were greater than a thousand-fold, with benzoic acid, caffeine, and dinitrochlorobenzene showing the greatest rates of permeation. Closely related compounds showed great differences in permeation. Benzoic acid was absorbed at 200 times the amount of its more water-soluble glycine conjugate, hippuric acid, thus confirming the premise that lipophilic compounds permeate more rapidly than hydrophilic compounds.

Sato and Nakajima measured the concentration of trichloroethylene in breath, blood, and urine after exposure by immersion of one hand in liquid solvent for 30 minutes.[41] Comparison with a four-hour exposure via inhalation at 100 ppm demonstrated a similar time course for excretion, although exposure times and concentrations were different. They postulated that trichloroethylene partitions in skin as a lipid-soluble material and is slowly released from the skin over a longer time course. Sato and Nakajima concluded that under normal industrial uses, the risk from skin absorption would be limited because of the pain associated with prolonged liquid exposure to trichloroethylene.

Several studies have reported on the percutaneous absorption of xylene in man. Using volunteers, Engstrom et al. monitored xylene concentrations (as the metabolite methylhippuric acid) in blood, exhaled air, and urine after immersion of both hands for 15 minutes.[42] They estimated that 35 mg of xylene was absorbed through the skin, which was equivalent to the quantity absorbed by the lung at 100 ppm for the same exposure duration. The absorption rate was 2 $\mu g/cm^2/min$. Riihimake and Pfaffli exposed volunteers to vapors of xylene, styrene, toluene, 1,1,1-trichloroethane, and tetrachloroethylene at 600 ppm for 3.5 hours.[2] Permeation was greater for xylene, styrene, toluene, and tetrachloroethylene than for 1,1,1-trichloroethane. Assuming a total body surface area for skin exposure, the authors estimated these exposures correspond to an exposure by inhalation of less than 10 ppm for 3.5 hours. The authors also demonstrated that mixtures of solvents influenced percutaneous absorption.[43] Direct exposure to a 50% mixture of xylene: isobutanol reduced the amount of xylene absorbed. When a water-saturated isobutanol solution was mixed with xylene, the quantities of xylene absorbed were comparable to that of pure xylene. Riihimake suggested the enhanced absorption with the aqueous mixture may have been a result of increased permeability caused by skin hydration.

***In Vivo* Animal Studies.** Permeation of a series of aliphatic chlorinated compounds was evaluated by topical application to abdominal skin of mice.[44] Among the solvents tested, those with the highest solubility in water had the highest rates of absorption. This observation is not consistent with predictions from skin permeation models[7] or other studies measuring percutaneous absorption in humans. Permeation rates increased linearly with concentration for each solvent. A 53-fold difference in permeation rate was demonstrated, with dichloromethane permeating most rapidly, followed by 1,2-dichloroethane, trichloro-methane, 1,1,2-trichloromethane, 1,1,2,2-tetrachloroethane, tetrachloromethane, 1,1,1-trichloroethane, and tetrachloroethylene. Permeation rates from these studies are similar to those calculated by Stewart and Dodd[38] and by Dutkiewicz and Tyras[40] in humans. The authors concluded that substantial risk from dermal permeation exists when the skin is exposed directly to liquid solvents.

Hefner et al. used monkeys to evaluate the permeation of vinyl chloride.[45] Whole-body exposures to atmospheres of 7000 and 8000 ppm vinyl chloride in a closed system for 2.5 hours demonstrated that 0.023 and 0.031% of the total vinyl chloride available was absorbed. Ignoring variables such as hair or skin composition, the authors estimated that skin absorption to an exposure of 7000 ppm was equivalent to an exposure via inhalation of 0.2 ppm for eight hours.

Using guinea pigs, Jakobson et al.[46] measured the blood levels of ten solvents after dermal application. Three patterns of kinetic absorption were observed. For relatively lipophilic compounds, blood concentrations increased rapidly in the first hour, then decreased for the remainder of the exposure. The authors suggested that either local vasoconstriction, or rapid transport from blood to fat, or biotransformation may account for this pattern. For benzene and 1,2-dibromoethane, blood levels remained constant after maximum levels were achieved, suggesting equilibrium with elimination had been achieved. More hydrophilic solvents, 1,2-dichloroethane and 1,1,2,2-tetrachloroethane, increased steadily in the blood throughout the exposure, suggesting that elimination was not in equilibrium with percutaneous uptake. Comparison of permeation of 1,1,1-trichloroethane with the more-water-soluble isomer 1,1,2-trichloroethane demonstrated twice the uptake of the isomer and suggested partitioning of the more lipophilic compound in the skin with limited release to the blood.

McDougal et al. evaluated the pharmacokinetics of organic vapor absorption using whole-body exposure of rats.[3] Animals with respirators were exposed for four hours to varying concentrations of nine different compounds; blood concentrations were then measured to determine permeability. Permeability correlated well with the fat:air partition coefficient and the muscle:air partition coefficient. Molecular weight of the solvent did not correlate with permeability. Permeability was greatest for styrene, xylene, and dibromomethane and less for bromochloromethane, toluene, benzene, methylene chloride, hexane, and isoflurane. The authors concluded that the importance of the dermal route of exposure is chemical-specific, and that permeation of lipophilic chemicals may contribute significantly to the total exposure. Contrary to other studies, they concluded that exposures to solvent vapors represent a substantial risk for dermal permeation.

In conclusion, these studies demonstrate that skin exposure to solvents in liquid form poses a substantial risk for dermal absorption,[38,39,41,42] but that the risks are much less, perhaps negligible, when skin is exposed to vapors.[2,45] They also demonstrated that the rates of skin penetration are highly variable and are dependent on the nature of the penetrant as well as the location, length, and duration of exposure.

Guidelines for Assessing the Risk for Dermal Permeation

While skin is normally an effective barrier against the permeation of many agents, certain occupational and environmental exposures may represent a risk for systemic toxicity. Factors that influence the risk for exposure by skin permeation are more numerous than those influencing inhalation exposures; consequently, the risks associated with skin exposure are less well defined. The most relevant information for assessing the risk from skin permeation would be data from actual workplace exposures and the supporting information necessary to substantiate the cause-effect relationship. This would include data from epidemiological studies, clinical reports of toxic effects with case histories, and prospective biological monitoring of a workforce. Unfortunately, such conclusive data are seldom available, and, given the current emphasis of occupational medicine and epidemiology, it is unlikely that such data will be available on a routine basis in the future.

In contrast, skin permeation studies have been frequently used to describe the penetration of chemicals in both humans and animals, and such studies may eventually provide a reliable means to rank chemicals according to their potential risk for systemic toxicity. The determination that a chemical will penetrate the skin does not establish the risk for systemic toxicity—only that the potential exists and that further data are needed to describe the pharmacodynamics of the penetrant. It will be difficult to test all chemicals by labor-intensive bioassays because of cost and time, and to date there is no reliable basis for extrapolation from one chemical to another.

In the absence of human-exposure data or skin-permeation studies, end-point biological tests or pharmacodynamic evaluations may provide a means for assessing the risk from skin permeation. Lethality, as determined by skin toxicity tests, is not a measure of skin permeation, but indicates the toxicity of a compound. Pharmacodynamic evaluations provide information about the metabolism, distribution, and excretion of a compound. As with skin penetration studies, not all chemicals can be evaluated, and the basis for extrapolation to chemicals of similar structure is not fully reliable. As data describing the penetration, metabolism, systemic distribution, toxicity, and excretion of different compounds become available, models may be developed to predict the risk of systemic toxicity from dermal exposure to a wide variety of compounds. However, until accurate models of skin permeation are developed, the health and safety professional must rely on available data from each of the above sources to estimate the risk for dermal absorption and apply appropriate protective measures.

References

1. US Bureau of Labor Statistics, *Annual Survey*, Washington, DC (1984).

2. V. Riihimaki and P. Pfaffi, "Percutaneous Absorption of Solvent Vapors in Man," *Scand. J. Work Environ. Health* **4**, pp. 73–85 (1978).

3. J.N. McDougal, G.W. Jepson, H.J. Clewell, III, and M.G. Anderson, "Pharmacokinetics of Organic Vapor Absorption," *Pharmacol. Ski*, **1**, pp. 245–251 (1987).

4. S.B. Tucker and M.M. Key, "Occupational Skin Disease" in *Environmental and Occupational Medicine*, W.N. Rom, Ed. (Little, Brown and Company, Boston, 1983).

5. Y.T. Whitton and J.D. Fuerall, "The Thickness of the Epidermis," *Brit. J. Dermatology* **89**, pp. 467–476 (1973).

6. E.L. Rangane, "Skin Structure, Function and Biochemistry," in *Dermatotoxicology*, F. N. Marzulli and H.I. Maibach, Eds. (McGraw-Hill, New York, 1983).

7. R.J. Scheuplein and I.H. Blank, "Permeability of the Skin," *Physiological Reviews* **51**, pp. 702–747 (1971).

8. A. Pannatier, B. Jenner, and B. Testa, "The Skin as a Drug Metabolizing Organ," *Drug Metab. Rev.* **8**, pp. 319–343 (1978).

9. H. Mukhtar and D.R. Bickers, "Drug Metabolism in Skin," *Drug Metabl. Disposit.* **9**, p. 311 (1981).

10. C. Fox, J. Selkirk, F. Price, R. Croy, K. Sanford, and M. Fox, "Metabolism of Benzo(a)pyrene by Human Epithelial Cells *in vitro*," *Cancer Res.* **35**, pp. 3551–3557 (1975).

11. R.J. Scheuplein, "Permeability of the Skin: A Review of Major Concepts," *Curr. Probl. Dermatol.* **7**, pp. 172–186 (1978).

12. P.M. Elias, K.R. Feingold, G.K. Meron, S. Grayson, M.L. Williams, and G. Grubaver, "The Stratum Corneum Two-Compartment Model and Its Functional Implications," *Pharmacol. Skin* **1**, pp. 1–19 (1987).

13. I.H. Blank, "Measurement of pH of the Skin Surface II—pH of the Exposed Surfaces of Adults with No Apparent Skin Lesions," *J. Invest. Dermat.* **2**, pp. 75–79 (1939).

14. R.R. Suskind, "Environment and the Skin," *Environ. Health Persp.* **20**, pp. 27–37 (1977).

15. E.A. Emmett, "Photobiologic Effects," in *Occupational and Industrial Dermatology*, H. I. Maibach, Ed. (Year Book Medical Publishers, Inc., Chicago, 1987).

16. R.T. Tregear, "Relative Penetrability of Hair Follicles and Epidermis," *J. Physiol.* **156**, pp. 307–313 (1961).

17. R.J. Feldmann and H.I. Maibach, "Regional Variation in Percutaneous Penetration of ^{14}C Cortisol in Man," *J. Invest. Dermatol.* **48**, pp. 181–183 (1967).

18. I.H. Blank, R.J. Scheuplein, and D.J. Macfarlane, "Mechanism of Percutaneous Absorption III—The Effect of Temperature on the Transport of Non-Electrolytes Across the Skin," *J. Invest Dermatol.* **49**, pp. 582–589 (1967).

19. K. Grice, H. Sattar, H. Baker, and M. Sharratt, "The Relationship of Transdermal Water Loss to Skin Temperature in Psoriasis and Eczema," *J. Invest. Dermatol.* **64,** pp. 313–315 (1975).

20. A. Bjornber, "Irritant Dermatitis," *Occupational and Industrial Dermatology,* H.I. Maibach, Ed., (Year Book Medical Publishers, Inc., Chicago, 1987).

21. G. Dupis and C. Benezra, *Allergic Contact Dermatitis to Simple Chemicals* (Dekker, Basel, 1982).

22. J.R.T. Reeves and H.I. Maibach, "Drug and Chemical Induced Hair Loss," in *Dermatoxicology and Pharmacology Advances in Modern Toxicology,* F.N. Marzulli and H.I. Maibach, Eds. (John Wiley & Sons, New York) **4,** pp. 487–500 (1977).

23. K.D. Crown, "Chloracne," *Seminars in Dermatology* **1,** pp. 305–314 (1982).

24. S.L. Gunpart, M.N. Honis, D.F. Roses, and A.W. Kopf, "The Dianosis and Management of Common Skin Cancers," *CA Cancer Journal for Clinician* **31**(2), pp. 79–90 (1981).

25. R.B. Fountain, "Occupational Melanoderma," *Br. J. Dermatol.* **79,** pp. 59–60 (1967).

26. E. Oliver, L. Schwartz, and L.H. Warren, "Occupational Leukaderma, Preliminary Report," *JANS* **113,** pp. 927–928 (1939).

27. C.D. Calnan, "Occupational Leukoderma from Alkyl Phenols," *Prec. R. Soc. Med.* **66,** pp. 258–260 (1973).

28. A.S. Michaels, S.K. Chardrosekoran, and J.E. Shaw, "Drug Permeation Through Human Skin: Theory and *in-vitro* Experimental Measurement," *AICHE Journal* **21,** pp. 985–86 (1985).

29. W.J. Albery and J. Hodgraft, "Percutaneous Absorption: Theoretical Description," *J. Pharmacol.* **31,** pp. 129–39 (1979).

30. H.Y. Ando, T. Schultz, R.L. Schnaare, and E.T. Sugita, "Percutaneous Absorption: A New Physicochemical Predictive Model for Maxiumum Human *in-vivo* Penetration Rates," *J. Pharm. Sci.* **73,** pp. 461–467 (1984).

31. B.W. Barry, "Penetration Enhancers: Made of Action in Human Skin," *Pharmacol. Skin* **1,** pp. 121–137 (1987).

32. W. Meyers, R. Schwarz, and K. Neward, "The Skin of Domestic Mammals as a Model for the Human Skin, with Special Reference to the Domestic Pig," in *Current Problems in Dermatology,* Vol. 7, G.S. Simon, Z. Poster, M.A. Klingberg, and M. Kaye, Eds., (Karger, Basel, Switzerland, 1978) pp. 39–52.

33. W.G. Reifenrath and G.S. Hawkins, "The Weaning Yorkshire Pig as an Animal Model for Measuring Percutaneous Penetration" in *Swine in Biomedical Research,* Vol. 1, M.E. Tumbleson, Ed. (Plenum Publishing Corp., New York, 1986).

34. J.P. Skelly, et al., "FDA and AAPS Report of the Workshop on Principles and Practices of *In Vitro* Percutaneous Penetration Studies: Relevance to Bioavailability and Bioequivalence," *Pharmaceutical Research,* **4**:3, pp. 255–267 (1987).

35. R.J. Feldmann and H.I. Maibach, "Percutaneous Penetration of Steroids in Man," *J. Invest. Dermatol.* **52,** pp. 89–94 (1969).

36. R.J. Feldmann and H.I. Maibach, "Absorption of Some Organic Compounds Through the Skin in Man," *J. Invest. Dermatol.* **52,** pp. 339–404 (1969).

37. R.J. Feldmann and H.I. Maibach, "Percutaneous Penetration of Some Pesticides and Herbicides in Man," *Toxicol. Appl. Pharmacol.* **28,** pp. 126–132 (1974).

38. R.D. Stewart and H.C. Dodd, "Absorption of Carbon Tetrachloride, Trichlorethylene, Tetrachloroethylene, Methylenechloride, and 1,1,1-Trichloroethane Through the Human Skin," *Am. Ind. Hyg. Assoc. J.*, pp. 439–446 (1964).

39. T. Dutkiewicz and H. Tyras, "A Study of the Skin Absorption of Ethylbenzene in Man," *Br. J. Indust. Med.* **24,** pp. 330–332 (1967).

40. T. Dutkiewicz and H. Tyras, "Skin Absorption of Toluene Styrene and Xylene by Man," *Br. J. Indust. Med.* **25:**3, p. 243 (1968).

41. A. Sato and T. Nakajima, "Differences Following Skin or Inhalation Exposure in the Absorption and Excretion Kinetics of Tricholoroethylene and Toluene," *Br. J. Indust. Med.* **35,** pp. 43–49 (1978).

42. K. Engstrom, K. Husman, and V. Riihimaki, "Percutaneous Absorption of M-Xylene in Man," *Int. Arch. Occup. Environ. Health* **39,** pp. 181–189 (1977).

43. V. Riihimaki, "Percutaneous Absorption of M-Xylene from a Mixture of M-Xylene and Isobutyl Alcohol in Man," *Scan. J. Work Environ. Health* **5,** pp. 143–150 (1979).

44. H. Tsuruta, "Percutaneous Absorption of Organic Solvents 1—Comparitive Study of the *In Vivo* Percutaneous Absorption of Chlorinated Solvents in Mice," *Indust. Health* **13,** pp. 227–236 (1975).

45. R.E. Hefner, P.G. Watanabe, and P.J. Gehring, "Percutaneous Absorption of Vinyl Chloride," *Toxicol. Appl. Pharmacol.* **34,** pp. 529–532 (1975).

46. I. Jacobson, J.E. Wahlberg, B. Holmberg, and G. Johansson, "Uptake Via the Blood and Elimination of Ten Organic Vapor Solvents Following Epicutaneous Exposure in the Anesthetized Guinea Pig," *Toxicol. Appl. Pharm.* **63,** pp. 181–187 (1982).

Chapter 3

Chemistry of Chemical Protective Clothing Formulations

Jimmy L. Perkins

Introduction

Many types of polymers are used for protective clothing. No two products are identical, though they may be of the same generic type. For example, nitrile gloves made by two different manufacturers will have somewhat different constituents (i.e., "recipes"). The way a polymer product is made has very important consequences to its protective properties; only slight changes in the manufacturing process can change these properties considerably. Some of the major variations in polymer systems include the use of copolymers, laminates, and blends. Also, the degree of crosslinking, crystallinity, and the amount of plasticizer and fillers in a polymer greatly affect the properties of the final product. This chapter will discuss the above variables, their chemistry, and how they affect the structure of the polymer product, which in turn will affect its protective properties. Further information is available in many books that deal with the science of polymer chemistry. This chapter is taken largely from the texts listed in the suggested readings at the end.

Polymer Types

Polymers are made from monomers, or single molecules bonded together in chains through some type of chemical reaction. The details of the chemical reactions are not important here, but a few examples follow.

Condensation is one of the most important polymer-forming reactions and is used in the manufacture of many polymers, including polyesters. A polyester is the condensation product of a dialcohol and a diacid, as illustrated in the following example:

$$\mathrm{R{-}OH} + \mathrm{HO{-}\overset{\overset{\displaystyle O}{\|}}{C}{-}R} \rightarrow \mathrm{R{-}O{-}\overset{\overset{\displaystyle O}{\|}}{C}{-}R'} + H_2O \qquad (3.1)$$

Alcohol + Acid → Ester + Water

The hydrogen from the alcohol and an OH group from the acid leave to form water and an acid ester. The alcohol and the acid must be bifunctional (reactive in two places) so that once the initial reaction occurs, the ends of the new product also may react to continue the process.

Another important polymer reaction is the *addition* reaction. In this reaction no molecule is split out, which means that the repeating unit has the same composition as the original monomer (unlike the condensation reaction, where the repeating unit is somewhat different from that of the monomer[s] that were used). In addition reactions, the important step is the opening of a double bond, as in the following example for polyetheylene made from ethylene.

$$\begin{array}{ccccccccccc} H & & H & & H & & H & & & H & & H & & H & & H \\ | & & | & & | & & | & & & | & & | & & | & & | \\ C & = & C & + & C & = & C & \rightarrow & - & C & - & C & - & C & - & C & - \\ | & & | & & | & & | & & & | & & | & & | & & | \\ H & & H & & H & & H & & & H & & H & & H & & H \end{array} \qquad (3.2)$$

In the simplest reaction to form this polymer, no hydrogen is lost from the ethylene groups; only the double bond is opened, leaving a free radical to react with other monomers. The same process is used in the manufacture of polyvinylchloride and other polymers made from double-bonded monomers including styrene, propylene, and acrylonitrile. Appendix A lists polymer types and structures used in protective clothing.

When two differing monomers react in a random sequence, a *copolymer* is formed. A random copolymer does not have a specific order to its constituent monomers. Appendix A includes some examples of copolymers used in protective clothing, such as Viton (a copolymer of hexafluoropropylene and tetrafluoroethylene), and nitrile rubber (a copolymer of acrylonitrile and butadiene). In both cases, the two types of molecules that form the polymer are found randomly throughout the polymer chain. The condensation example above (3.1) may appear to yield a copolymer. However, even though two distinct molecules are used in making the polymer, these molecules must be arranged in a specific order for the condensation reaction to occur, i.e., there is only one repeating unit.

Figure 3-1 shows examples of polymer types. In some cases *block copolymers*, as opposed to random copolymers, are formed. A polymer *blend* is a mixture of two polymer resins, not a copolymer. To complete the analogy above, in a blend the **A** monomers would lie next to the **B** monomers, but would not be attached by chemical bonds, as in a block copolymer. A *laminate* is simply a multilayer of two or more polymer sheets that are adhered together.

A *linear* polymer has no side chains. For example, polyethylene, which forms a somewhat zigzag pattern, is nevertheless a linear polymer because (in its simplest form) there are no side chains attached to the polymer strand. *Branched* polymers have side chains appearing along the backbone of the main polymer strand. These side chains develop because of the occasional reaction of the polymer backbone with monomer molecules at locations other than the ends. If a trifunctional (reactive in three places) monomer is introduced in an exact place or at random along a linear chain, then two of its functional groups will be taken up as the polymerization process proceeds along the backbone chain, leaving one functional group to react to form a branch. These branching reactions can occur by chance, but in some cases a certain amount of trifunctional monomer is added to a polymer batch to induce branching. Branching has a tremendous influence on the properties of polymers.

The third type of polymer arrangement is the *crosslinked*, or networked, polymer, which is an extreme case of a branched polymer (Figure 3-1). As the length and frequency of branches on a polymer chain increase, the branches will eventually reach from chain to chain and become crosslinked. As the chains become more and more connected in three dimensions, the entire object becomes one huge molecule. As an extreme example, a bowling ball is one large network polymer.

Crosslinks may be formed by adding sufficient amounts of trifunctional monomers. Crosslinks may also be formed using the process known as "curing." Charles Goodyear first used sulfur to cure natural rubber. His curing process caused the natural rubber to harden because the added sulfur linked the polymer chains together, decreasing their slipping and sliding against one another. Without this curing process, natural rubber rapidly loses its shape and becomes a sticky liquid at room temperature.

A cured polymer becomes hard and cannot be melted without forcing a decomposition of the crosslinked structure. This type of polymer is known as *thermosetting*. To decompose a thermosetting polymer requires extended exposures to heat or

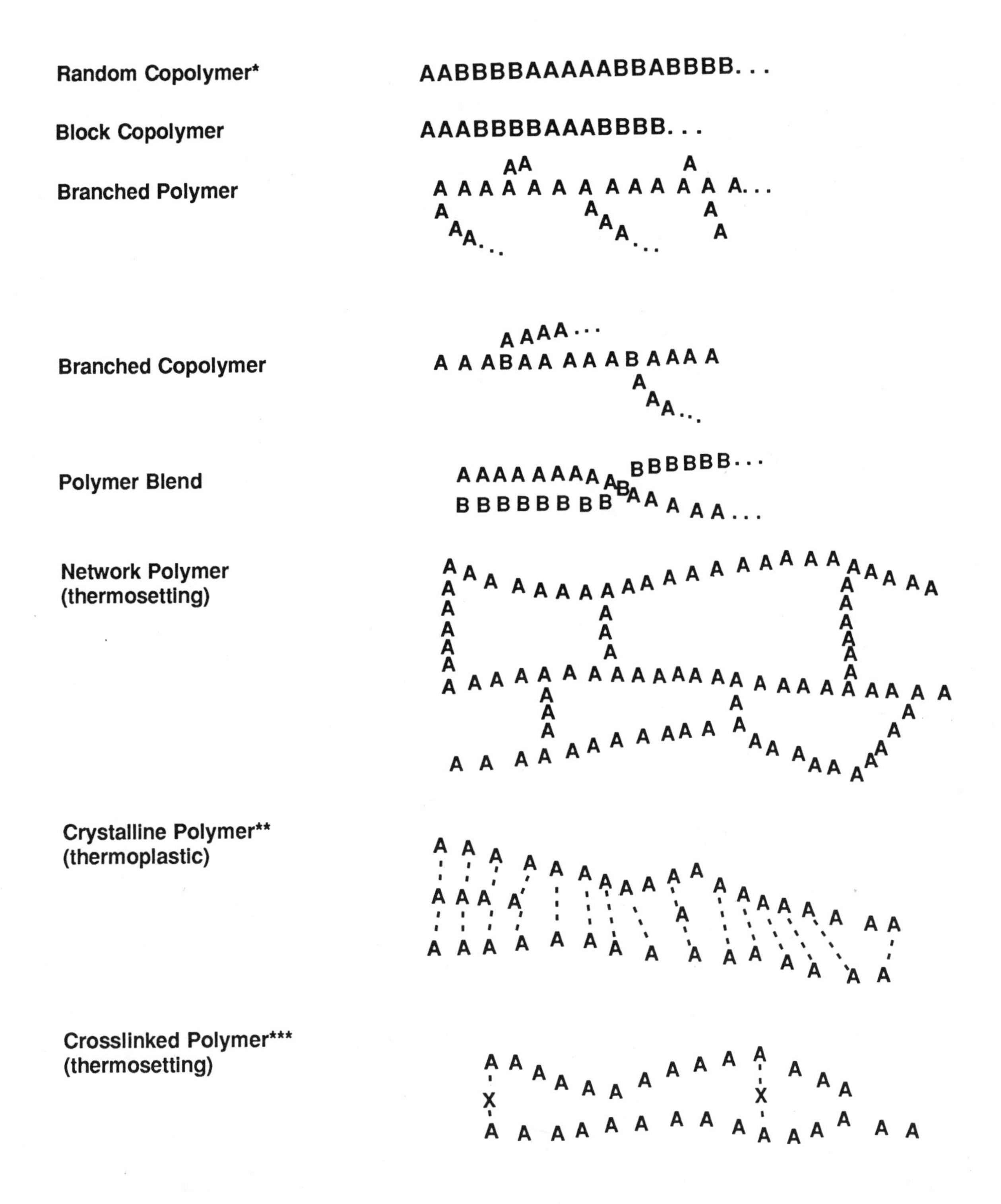

*A, B are different repeating units
**- - represents molecular interactions
***X represents a crosslink

Figure 3-1. Types of polymer structures.

chemical reagents; this decomposition is not reversible. By analogy, an egg is a thermosetting substance.

Other polymers are *thermoplastic*, meaning that after melting they reversibly regain their solid nature when cooled again. Chocolate is an example of a thermoplastic substance. In thermoplastic polymers, the energy holding the polymer in a solid or rubbery form is due to molecular interactions rather than crosslinks. These molecular interactions are also important in a polymer's degree of crystallinity. Molecular interactions and crystallinity will be discussed later.

A few other terms of polymer nomenclature should be discussed. *Plastic* substances can be shaped or molded and will hold their forms at constant temperatures; waxes and moist clay are examples of this state. (Polyvinylchloride is a plastic, but through the addition of plasticizers it can be made flexible.) A *rubber* substance, an *elastomer*, has the additional ability to withstand and recover from deformation or elongation under stress. The difference between a plastic and an elastomer is partly a function of the crosslinking between the polymer strands. This can be affected by the introduction of additives (plasticizers) that change the bonding in the polymer (discussed in the next section). For the most part, face-shield materials and fibers used in chemical protective clothing are plastic substances. Substances (usually thermosetting) used to coat fibers or form gloves are elastomeric.

One final differentiation is the method by which protective clothing is manufactured. Solvent dipping, milling, and latex-suspension methods are all very important in determining the characteristics of the final product. These manufacturing processes are discussed in Chapter 6, and their effects on permeation characteristics are discussed in Chapter 4.

Polymer Chemistry

In order to see the characteristics that make a polymer good for protective clothing, one must have an understanding of polymer structure and the various types of chemical bonds holding polymers together. Table 3-1 lists the different types of bonds and the amount of energy necessary to break them. Carbon atoms form *covalent* (or primary) bonds with one another and with several other elements. As can be seen in Table 3-1, these bonds are quite strong, with a relatively large amount of disassociation energy required to break the bonds. The monomers in a polymer are generally held together by covalent bonds, as are the links in a crosslinked polymer (although the latter in some cases are held together by ionic bonds). For the purpose of understanding important CPC characteristics, the covalent bond is the most important bond in the polymer backbone and in the crosslinking of polymers. The three remaining bonds—dispersion forces, polar bonds or dipole interactions, and hydrogen bonds—are all molecular interactions, attractions, or secondary bonds, and they are not generally thought of as "bonds" in the true chemical sense. As can be seen in the table, the disassociation energies of these three types are much smaller than those for covalent bonds. These molecular interactions represent the cohesive forces responsible for holding molecules together in a liquid. *Dispersion* forces are solely responsible for holding together non-polar-bonding or non-hydrogen-bonding liquids, such as pentane, hexane, and other straight-chain hydrocarbons. Dispersion forces arise from the temporary dipole moments that fluctuate about a molecule, depending on positional shifts in its electron cloud. *Polar* bonding is present between many molecules that contain oxygen, fluorine, or nitrogen atoms; for example, ketones, fluorinated hydrocarbons, and amines. Polar bonding is due to the permanent dipoles present in these molecules.

Table 3-1. Types of bonds and their disassociation energies.

Type	Disassocation energy (kcal/mole)
Covalent	50–200
Ionic	10–20
Hydrogen	3–7
Polar	1.5–3
Dispersion	0.5–2

Adapted from Rosen.

Hydrogen bonding is a special case of polar bonding, present in molecules that contain two polar ends, one of which is an exposed hydrogen atom. The best example of hydrogen bonding occurs in water where hydrogen from one ion readily interacts with the oxygen from another. Hydrogen bonding is also very important in alcohols and organic acids. Examples of these three molecular interactions are shown in Figure 3-2.

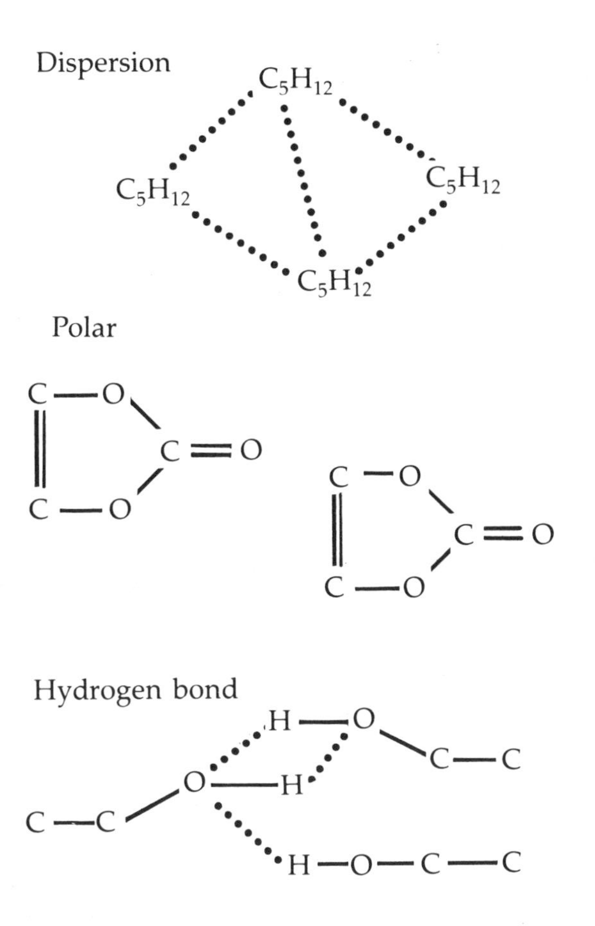

Figure 3-2. Types of molecular interactions or secondary bonds.

The various kinds of bonds determine the final structure of the polymer. If the polymer is not cured or otherwise crosslinked, only secondary bonds hold the polymer together. If the temperature is raised, the molecules become more volatile, just as the molecules become more volatile when the temperature is raised in a liquid. Consequently, a solid polymer in which the chains are held together only by secondary bonds (a thermoplastic polymer) at room temperature will begin to dissociate and melt as the temperature is raised. On the other hand, a thermosetting polymer has covalent bonding between the polymer chains.

This covalent bonding is usually provided by a crosslinking agent (an accelerator) or by natural crosslinking between the polymer chains due to the presence of trifunctional monomers. A thermosetting polymer will not melt until it reaches its *degradation temperature,* the temperature at which many of the covalent bonds break; at this temperature the polymer has degraded into smaller molecules. It requires considerably more energy typically to degrade a crosslinked polymer than to melt a thermoplastic polymer. Molecular interactions and covalent bonding or crosslinking are extremely important in terms of the ability of foreign molecules to permeate through polymers.

A high degree of molecular interaction leads to greater *crystallinity* of the polymer. Although no polymers are purely crystalline, some such as high-density polyethylene approach this state. Crystalline polymers are a metal-like lattice, and few spaces are left for permeant molecules to pass between the polymer chains. Crystallinity can be induced in a polymer by stretching it as it cools; this effectively helps to align the chains, increasing the opportunity for chain interaction. Crosslinking can also inhibit the passage of permeant molecules, since the polymer chains are held together by much stronger forces than in crystallinity. In both crystallinity and crosslinking, not only are the spaces between polymer chains made much smaller, but the random movement of the polymer chains is also restricted. Consequently, the probability for holes of a certain size to open between the polymer chains becomes lower, and permeation is reduced. As will be discussed in Chapter 4, permeation is a function of both solubility and diffusivity.

Solution of the permeant molecules into the polymer via molecular interactions is important to the permeation process. In a crosslinked polymer, the polymer-solvent secondary bonds cannot overcome the polymer-polymer covalent crosslink bonds; consequently, crosslinked polymers generally are not soluble. When an organic solvent is applied that will interact with these polymers, swelling occurs (sometimes to a very great extent), but the polymers usually will not decompose. On the other hand, non-crosslinked polymers will easily dissolve when the appropriate solvent is used. An appropriate solvent for such a polymer has secondary bonding energies very similar to those of the polymer repeating units.

Agents added to polymers in order to reduce crystallinity are known as *plasticizers*. Commonly used plasticizers are the esters of phthalic and adipic acid. Plasticizers form secondary bonds between the polymer chains and serve to spread the chains apart, which in effect increases their flexibility. Since the bonds between plasticizer and polymer are not covalent, the appropriate solvents can leach the plasticizer out of the material. When a plasticizer is dissolved out of polymer material, the polymer once again becomes hard and inflexible.

Another way of plasticizing a polymer is to incorporate the plasticizer directly into the polymer chain through covalent bonding. In these situations, a second monomer whose polymer is usually very soft, is used as a plasticizer. Small amounts of "soft" monomer are added to the recipe where they are incorporated at random positions into the polymer chain. These internal plasticizers cannot be removed by solvent.

Geometric differences in polymers can also affect permeation characteristics. Molecules having the same atoms in different geometric arrangements are called *isomers*. In polymers containing symmetric monomers, isomerism is not important; however, if the monomer is not symmetrical (such as vinyl chloride), then the specific positions of the chlorine atoms along the polymer chain can indeed be important and can change the properties of the polymer. In general, if the chains formed by the unsymmetrical monomer are somewhat regular in structure, then the degree of crystallinity or interaction between the chains is increased, leading to a harder, tougher material. On the other hand, if the chains are twisted or irregular, then a softer polymer is usually formed, due to the decrease of molecular interaction between the chains.

The molecular weight of a polymer can be important in terms of both its final form and its resistance to permeation. The molecular weight of a polymer is an average molecular weight based upon the probable distribution of chain lengths present. The important point pertaining to molecular weight is that, at the ends of each polymer chain, the end groups may have chemical characteristics different from the repeating units of the polymer chain. If the molecular weight is small, then the chain length is small, and therefore the concentration of the end groups is greater. If the end groups are polar, and the polymer repeating groups are non-polar, then a low molecular weight may increase the permeability of the polymer to polar solvents. By the same analogy, if interaction occurs between repeating groups but not between repeating groups and ending groups, then there will be less chain interaction (and less crystallinity) as molecular weight decreases. In general, the higher the molecular weight, the lower the solubility of a polymer.

Polymer Transitions

When cooled to a low enough temperature, any polymer will assume the characteristics of glass —hardness, stiffness, and brittleness. At this temperature, called the *glass-transition temperature*, it is thought that large-scale molecular motion does not take place; instead, only atoms and small groups of atoms are able to move against the local restraints of secondary bond forces. As the polymer is heated, it becomes plastic again at the transition temperature, and large-scale molecular motion begins once more.

While all polymers have a glass-transition temperature, in highly crystalline polymers the transition temperature is difficult to detect. In these polymers, a second transition—melting—is approached at about the same time as the glass transition. In highly crystalline polymers, this temperature is comparable to the melting point in low-molecular-weight compounds, i.e., the point at which sufficient heat is added to the system to overcome most molecular interactions due to secondary bonding. The melting temperature of a polymer decreases with degree of crystallinity. This mirrors the phenomenon observed in liquids when impurities are added; for example, the melting temperature of water is lowered when salt is added. By analogy, the amorphous or non-crystalline portion of the polymer may be considered an impurity.

After a polymer has been heated past its melting temperature, it undergoes a final transition as the polymer degrades, at the degradation temperature.

In summary, a polymer is glassy at low temperatures. With increasing heat, it becomes plastic or perhaps rubbery at the first transition; at the second transition, it melts or becomes liquid-like; and finally at the third transition it degrades. In polymer chemistry, the glass-transition temperature and the factors that influence it are extremely important in manufacturing a polymer that has certain desired qualities.

Free volume within the polymer is the amount of space the polymer occupies that is not actually filled by the polymer chains. The greater the free volume, the more room the molecules will have for displacement, and the lower the glass-transition temperature will be. Increased free volume also means more room for permeants to penetrate the polymer.

The mobility, or the freedom of polymer chains to rotate internally, also affects the glass-transition temperature. If the individual polymer chains have difficulty in rotating because pendant groups interfere with one another, then the glass-transition temperature will be increased. For chains to rotate freely, additional energy is often needed for the pendant groups to "pass by" one another during the rotation. If the energy requirement is too large and the rotation does not occur, then the glass-transition temperature is usually higher than for a polymer with similar structure (isomers or near isomers) that does not possess the rotation problem.

Crosslinking in small amounts does not appreciably affect the glass-transition temperature. A high degree of crosslinking, however, restricts chain mobility and thereby raises the glass-transition temperature.

Summary

As stated at the beginning of this chapter, many books have been written on polymer chemistry, and the subject is obviously much more complex than the brief review given here. The purpose here was to give an introduction to polymer chemistry, and to the various relationships that are necessary for a complete understanding of chemical permeation through protective clothing. The most important relationships are listed below.

1. Protective clothing materials may be pure polymers, copolymers, laminates of polymers, or blends of polymers.
   a. Pure polymers are pure in the sense that they contain only one repeating unit.
   b. Copolymers contain at least two repeating units, both of which contain two functional groups.
   c. Blended polymers are composed of two or more polymer chains that have been previously polymerized; these chains are mixed in the final manufacturing process to form a polymer product that is a blend of two or more polymer chain types.
   d. Laminates are cemented, or otherwise joined, polymer sheets made up of different individual polymer types.

2. Crosslinking and crystallinity are determined by the types of bonds in the polymer.
   a. Crosslinking is dependent upon the same type of bonding (covalent) that holds repeating units together in the backbone of a polymer.
   b. Crystallinity is dependent upon secondary bonding or molecular interactions between the monomers and the polymer strand.

3. Through secondary bonding, plasticizers interact with the polymer strands, holding them apart to increase the polymer's flexibility.

4. Polymer transitions are important in the manufacture of a final polymer product.
   a. The glass-transition temperature is the temperature below which a polymer is glassy-like. Above this temperature, the polymer is plastic if it contains a relatively high degree of crystallinity; it is rubbery if it is amorphous.
   b. The melting temperature is the point above which the polymer melts or becomes a liquid due to the breakage of the molecular interaction forces between the polymer strands.
   c. The degradation temperature is the temperature at which covalent bonds begin to break.

In the following chapter, each of these properties will be discussed as it pertains to the permeation process. Previous and future attempts at modeling permeation based on a knowledge of the above factors will also be discussed.

Suggested Reading

F.W. Billmeyer, *Textbook of Polymer Science*, 3rd ed. (Wiley Interscience, New York, 1984). [Complete reference, but more complicated than Rosen (below).]

J. Brandrup, and E.H. Immergut, Eds., *Polymer Handbook*, 2nd ed. (Wiley Interscience, New York, 1975). [Physical-chemical constants for polymers.]

H.G. Elias and R.A. Pethrick, *Polymer Yearbook*, 1st ed. (Harwood Academic, New York, 1984). [Yearly reviews of important subjects.]

S.L. Rosen, *Fundamental Principles of Polymeric Materials* (Wiley Interscience, New York, 1982). [Short, concise, easy to understand.]

R.B. Seymour, *Modern Plastics Technology* (Reston Publishing, Reston, VA, 1975). [Good reference for polymer structures.]

Appendix A
Common names and repeating units for polymers used in chemical protective clothing

Name (constituents)	Repeating Unit
Flexible Films	
Butyl (97% isobutylene/ 3% isoprene copolymer)	$\left[\left(-C(CH_3)_2-CH_2-\right)_x\left(-CH_2-C(CH_3)=CH-CH_2-\right)_y\right]_n$
Chlorobutyl (butyl rubber with chlorine atoms substituted randomly for hydrogens)	see Butyl
Chlorinated polyethylene (polyethylene with 36–45% by weight chlorine atoms substituted randomly for hydrogen atoms)	see Polyethylene
EVA/PE (blend of 14% ethylene/ vinyl acetate copolymer; 86% polyethylene	$\left[\left(-CH_2-CH_2-\right)_x\left(-CH_2-CH(O-C(=O)-CH_3)-\right)_y\right]_n;\ \left(-CH_2-CH_2-\right)_n$
EVOH (ethylene/vinyl alcohol copolymer)	$\left[\left(-CH_2-CH_2-\right)_x\left(-CH_2-CH(OH)-\right)_y\right]$
Fluorine/Chloroprene (Viton, chloroprene laminate)	see Viton, Neoprene
FEP, fluorinated ethylene propylene resin (Hexafluoropropylene/ tetrafluoroethylene copolymer	$\left[\left(-CF_2-CF(CF_3)-\right)_x\left(CF_2-CF_2-\right)_y\right]_n$
FEP/TFE (FEP; tetrafluoroethylene blend)	FEP; $\left(-CF_2-CF_2-\right)_n$
Natural rubber (isoprene)	$\left(-CH_2-C(CH_3)=CH-CH_2-\right)_n$

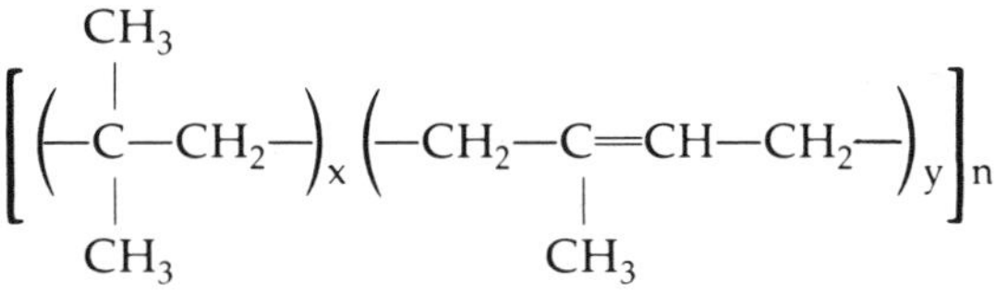

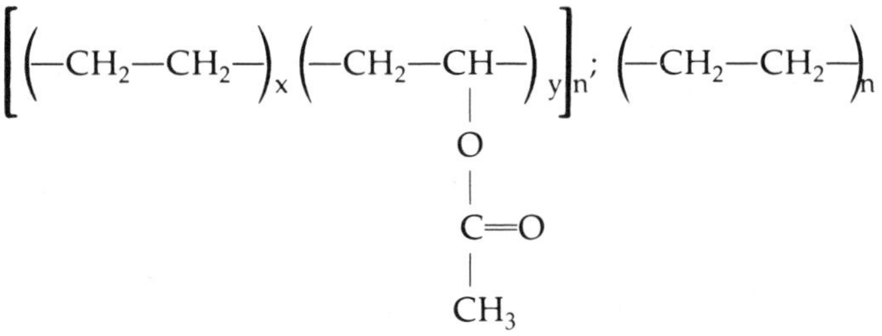

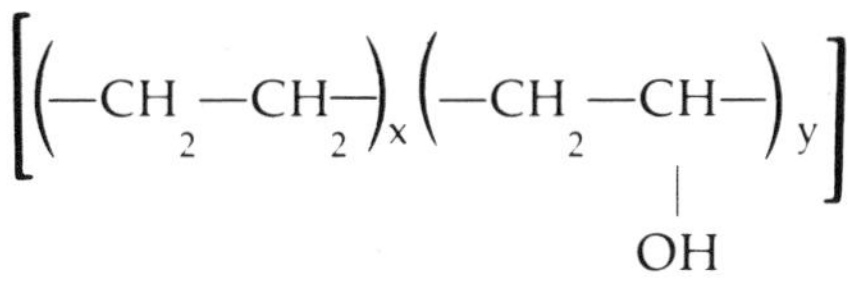

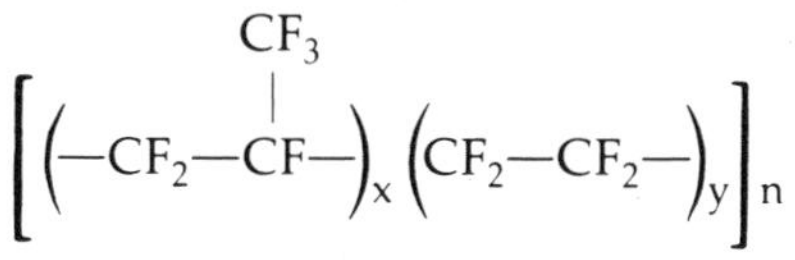

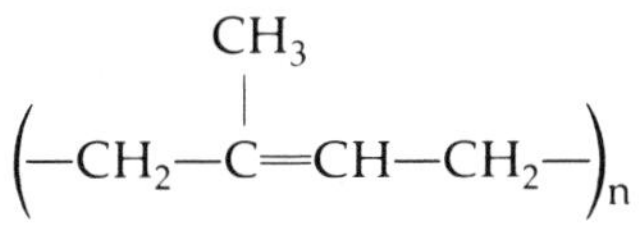

NBR, Nitrile (random or alternating acrylonitrile/butadiene copolymer	$\left[\left(-CH_2-\underset{}{\overset{CN}{\overset{\vert}{C}H}}-\right)_x\left(-CH_2-CH{=}CH-CH_2-\right)_y\right]$
Neoprene (chloroprene)	$\left(-CH_2-\overset{Cl}{\overset{\vert}{C}}{=}CH_2-CH_2-\right)_n$
Neoprene/SBR (chloroprene; styrene-butadiene blend)	See Neoprene; SBR
Neoprene/Natural (chloroprene; isoprene blend)	see Neoprene; Natural rubber
Polyethylene (ethylene)	$\left(-CH_2-CH_2-\right)_n$
PVA (vinyl alcohol)	$\left(-CH_2-\underset{OH}{\underset{\vert}{C}H}-\right)_n$
PVC (vinyl chloride)	$\left(-CH_2-\underset{Cl}{\underset{\vert}{C}H}-\right)_n$
PVC/Nitrile (blend)	see PVC; NBR
Saran (85% vinylidene chloride/15% vinyl chloride copolymer)	$\left[\left(-CH_2-\underset{Cl}{\underset{\vert}{\overset{Cl}{\overset{\vert}{C}}}}-\right)_x\left(-CH_2-\underset{Cl}{\underset{\vert}{C}H}-\right)_y\right]_n$
Saranex (laminate of polyethylene and saran)	see structures elsewhere
SBR (25% styrene/75% butadiene random copolymer)	$\left[\left(-CH_2-\underset{C_6H_5}{\underset{\vert}{C}H}-\right)_x\left(-CH_2-CH{=}CH-CH_2-\right)_y\right]_n$
Silver shield, 4H (laminates of polyethylene and EVOH)	see structures elsewhere
Viton (hexafluoropropylene/ vinylidene fluoride random copolymer)	$\left[\left(-CF_2-\underset{CF_3}{\underset{\vert}{C}F}-\right)_x\left(-CH_2-F_2-\right)_y\right]_n$
Vitrile (Viton; nitrile blend)	see Viton; NBR
Troionic (natural; neoprene; carboxylated nitrile blend	see Natural; Neoprene; NBR (carboxylated nitrile has carboxyl groups at the end of the polymer chains

Urethane (condensation product of a polyisocyanate and a polyol) — $(\text{—O—R—O—C(=O)—NH—R'—NH—C(=O)—})_n$

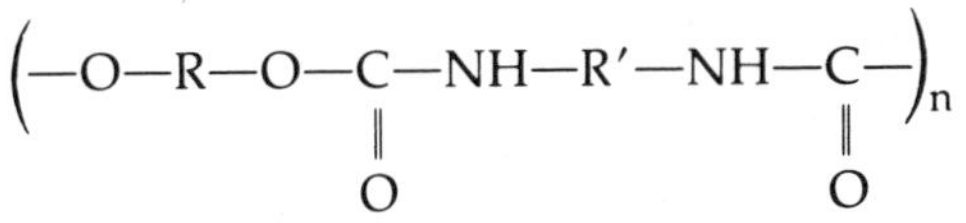

Commonly Coated Fabrics

Cellulose or cotton (polysaccharide, or polyalcohol)

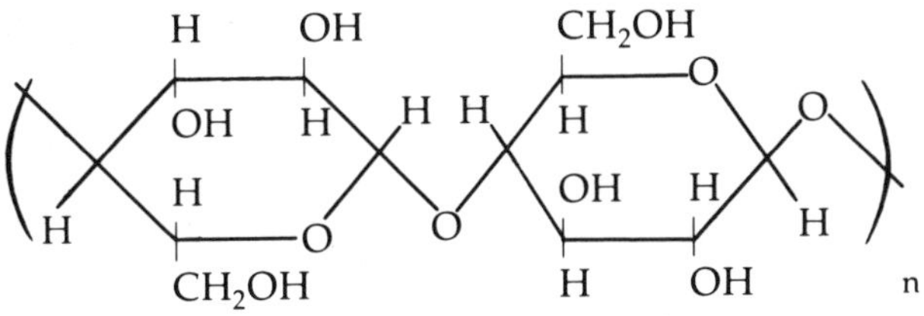

Disposaguard (cellulose reinforced with nylon, i.e., scrim reinforced) — see cellulose

Duraguard (non-woven polypropylene fibers) — $(\text{—CH}_2\text{—CH(CH}_3\text{)—})_n$

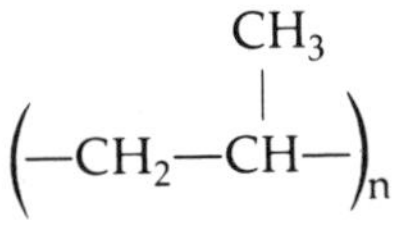

Fiberglass (fibers of glass, usually coated with silicon or silanes) — may be added to many polymers (polyester, nylon, polypropylene)

Gore-Tex (TFE; polyester or Nomex laminate) — see structures elsewhere

Nomex (aramid fiber, polyamide) — $(\text{—NH—C}_6\text{H}_4\text{—NH—C(=O)—C}_6\text{H}_4\text{—C(=O)—})_n$

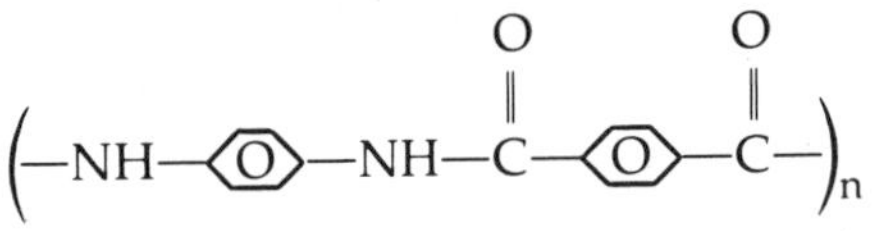

Nylon (polyamide, condensation product of diamine and dicarboxylic acid, e.g., nylon 66) — $(\text{—NH—R—NH—O—C(=O)—R'—C(=O)—O—})_n$

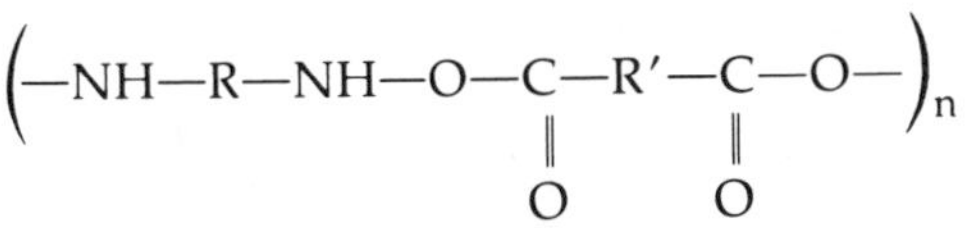

Polyester (condensation product of ethylene glycol and terepthalic acid), e.g., Dacron and Sontara — see under "Face Shield Polymers"

Rayon (regenerated cellulose fiber) — see cellulose

Safeguard (three layers of polypropylene) — see polypropylene

Tyvek (non-woven polyethylene fibers) — see polyethylene

Face Shield Polymers

Polymer	Structure
Acetate (cellulose acetate, acetate ester of cellulose).	see cellulose
Acrylic (methylmethacrylate)	$\left(-CH_2-\underset{\underset{\displaystyle OCH_3}{\mid}}{\underset{C=O}{\underset{\mid}{\overset{\overset{\displaystyle CH_3}{\mid}}{C}}}}-\right)_n$
Cellulose propionate (propionic ester of cellulose)	see cellulose
CR-39 (diethylene glycol bis allylcarbonate)	$\left[O\left(-CH_2-CH_2-O-\overset{\overset{\displaystyle O}{\parallel}}{C}-O-CH_2-CH(-CH_2-)\right)_2\right]_n$
Polycarbonate (condensation product of bisphenol-A and phosgene)	$\left(-O-C_6H_4-\underset{\underset{\displaystyle CH_3}{\mid}}{\overset{\overset{\displaystyle CH_3}{\mid}}{C}}-C_6H_4-O-\underset{\underset{\displaystyle O}{\parallel}}{C}-\right)_n$
Polyester (condensation product of dicarboxilic acid and glycol; for example terepthtalic acid and ethylene glycol)	$\left(-O-CH_2-CH_2-O-\underset{\underset{\displaystyle O}{\parallel}}{C}-C_6H_4-\underset{\underset{\displaystyle O}{\parallel}}{C}-\right)_n$
Polysulfone (condensation product of bis-phenol A and dichlorophenyl sulfone)	$\left(-O-C_6H_4-\underset{\underset{\displaystyle CH_3}{\mid}}{\overset{\overset{\displaystyle CH_3}{\mid}}{C}}-C_6H_4-O-C_6H_4-\underset{\underset{\displaystyle O}{\parallel}}{\overset{\overset{\displaystyle O}{\parallel}}{S}}-C_6H_4-\right)$
PVC (vinyl chloride)	$\left(-CH_2-\underset{\underset{\displaystyle Cl}{\mid}}{CH}-\right)_n$
Surlyn (ethylene/methacrylic acid, random copolymer	$\left[\left(-CH_2-CH_2-\right)_x\left(-CH_2-\underset{\underset{\displaystyle CH_3}{\mid}}{\overset{\overset{\displaystyle COO^-M^+}{\mid}}{C}}-\right)_y\right]_n$

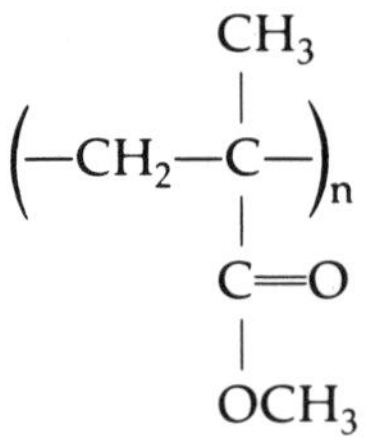

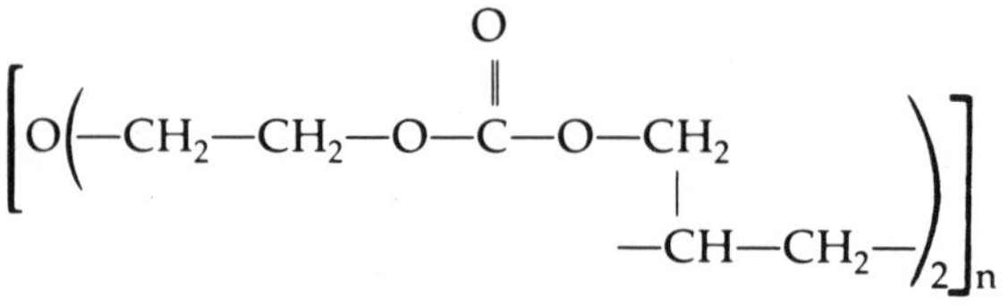

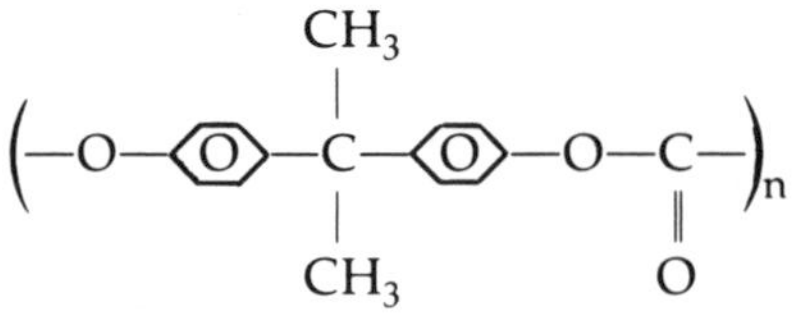

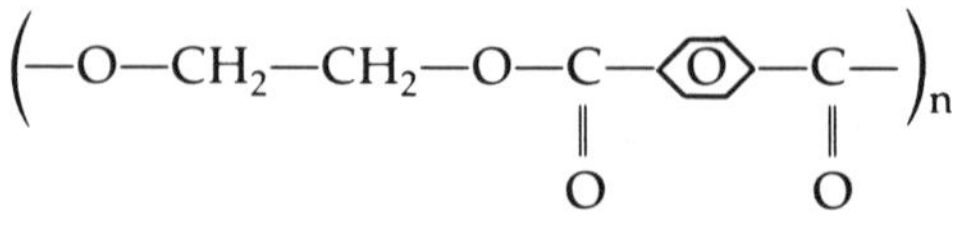

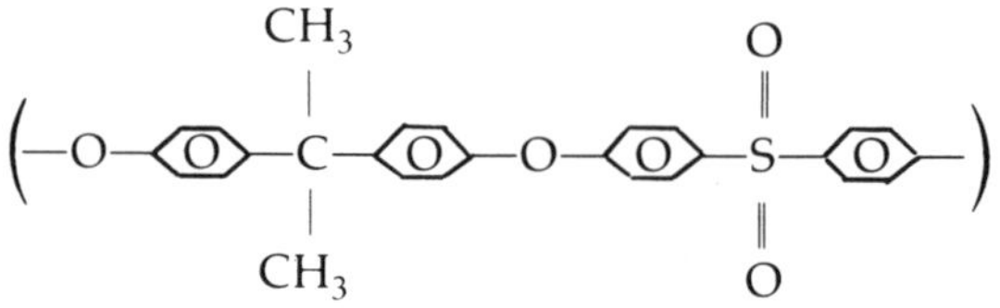

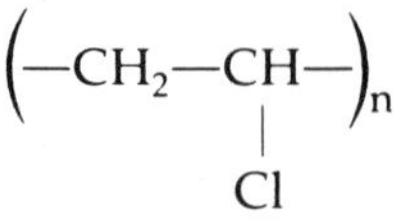

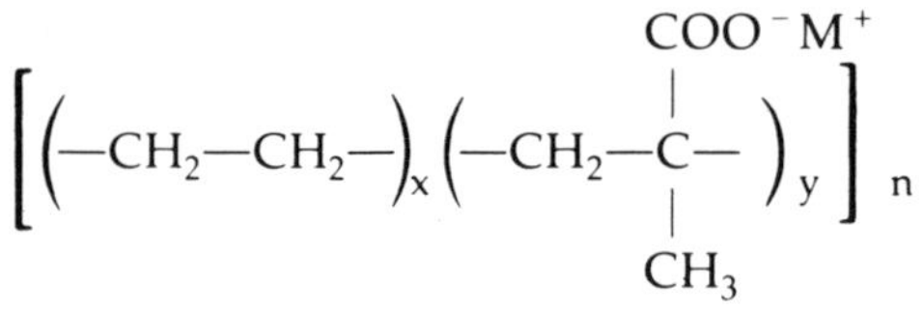

Chapter 4

Solvent-Polymer Interactions

Jimmy L. Perkins

Background

Although the major purpose of this chapter is to discuss solvent-polymer interactions as they relate to chemical protective clothing, the solutions of similar interaction problems must be mentioned first. These additional applications have longer histories in terms of research and use than protective clothing. For example, as one might expect from examining the vast array of plastic and elastomer products used in everyday life, there is a great interest in the possible effects that environmental contaminants may have upon a finished polymer product, such as how ozone can affect natural rubber products.[1]

Many polymer products are produced from a polymer resin dissolved in a solvent that later evaporates. In this chapter, "solvent" will be used to describe any non-polymer material of interest which contacts the polymer. The manner in which the solvent interacts with the polymer will largely determine the polymer drying time and the lasting quality of the product. Paints and coatings are perhaps the best examples of this application; once the paint film has formed, the remaining solvent must diffuse through the polymer and evaporate to the outside. The manner in which this occurs can affect the quality of the final product.[2]

The automotive industry encounters situations that are somewhat similar to protective clothing problems. Gaskets and seals made of polymer compounds must have a resistance to the various solvents and chemicals with which they come in contact. For example, the polymer parts of carburetors must have solvent resistance to gasoline.[3]

In other applications of polymers, the promotion of solvent-polymer interactions is used in order to accomplish a desired effect. For example, polymer films may be used to separate a mixture of solvents. In a particular mixture, one component may permeate more rapidly through the film than the other components, separating that component from the mixture.[4]

No polymer is indefinitely resistant to a chemical, and consequently *rates* are very important in solvent-polymer interactions. The rate variable is used in applications involving the timed-release concept. If an active ingredient (such as a pesticide or drug) is divided into small particles and encapsulated in polymer spheres of varying diameters or varying polymer types, then the active ingredient will be released at varying times. In many such situations, water is the solvent interacting with the polymer to release the active ingredient. Other applications involve the binding of the ingredient into the polymer matrix below the saturation point. Then the ingredient will diffuse through the polymer and evaporate to the environment at some determinable rate.[5]

The permeation process resulting from solvent-polymer interactions is used in several makes of passive dosimeters for sampling gases or vapors. By covering the dosimeter with a polymeric compound that is differentially permeable to a gas or vapor of interest, one may increase the selectivity of the device for that particular gas or vapor, and thus increase the passive dosimeter's accuracy.[6] Permeation-type dosimeters should not be confused with dosimeters that operate strictly on the principle of diffusion. Although some diffusion-type monitors have polymer membranes, their hole sizes are such that they do not interact with diffusing molecules (for example, the organic vapor monitors of the 3-M Co.).

In all the above examples, knowledge of solvent-polymer interactions is necessary to make useful predictions. As stated earlier, a given polymer is not indefinitely impermeable to a given chemical, although for certain solvents the time necessary for detection of a chemical traversing the thickness of the polymer can be on the order of days or weeks. Many ketones will permeate the polymer Viton, a fluorocarbon, in a matter of minutes; propanol, however, requires days before permeation can be detected.[7]

Given an understanding of the mechanisms involved in solvent-polymer interactions, then the relationships between the appropriate solvent and polymer physical/chemical properties should allow some prediction of the interaction parameters, such as the rate of permeation. While the chemical and physical properties involved are many and varied, the most important drawback to predictive modeling is the fact that protective clothing products are never pure polymers. They contain additives that change the chemical composition of the products and make the prediction process much more complicated.

The main purpose of this chapter is to discuss the mechanisms by which solvents and polymers interact, and also to discuss the previous attempts to predict or model these interactions. These previous modeling attempts show that qualitative predictions can be made about permeation rates and breakthrough times. This can be done using known physical/chemical data for the polymer and solvent. Most important, the accuracy of the predictions should improve, since the application of research to chemical protective clothing is relatively recent.

Solvent-Polymer Interaction Processes

Perhaps the best way to visualize the interaction between solvents and polymers is to imagine that the polymer sheet is made up of individual polymer molecule chains, almost like a bowl of spaghetti or a tangled mass of earthworms; in some situations the chains may be oriented in a more parallel fashion. Sometimes the chains are chemically (covalently) bonded at various points along their length. Through random motion, the chains open and close, making "holes" between them. These random movements of the chains, coupled with the random kinetic movement of the solvent molecules, underlie the process of the solvent permeating through the polymer.

Usually the solvent must first interact with the polymer chains before it can diffuse through the polymer. The magnitude of these interactions is dependent on the *cohesive energy* (the energy that holds liquid or solid molecules together) of the solvent and polymer. The cohesive energy for small molecules is related to vapor pressure—i.e., the more cohesive a solvent, then the lower its vapor pressure.

Compounds that have small cohesive energies are gases at room temperature.[8] According to solubility theory, two molecules with similar cohesive energies will have maximum solution in one another. Consequently, if a solvent and polymer have similar cohesive energies, then the likelihood that the solvent will interact with the polymer is great.[9]

This solvent-polymer interaction is the "solution step" in the permeation process. As shown in Figure 4-1, solvent molecules come in contact with the polymer and are dissolved depending on the interaction between solvent and polymer. Picture an initial "layer" of polymer molecules that the solvent would encounter; the number of solvent molecules entering that layer is dependent on solubility of the solvent in the polymer (i.e., the similarity of the two compounds), the size of the solvent molecules, and the size of holes opening in the polymer layer.

In order to enter the second layer, solvent molecules must move forward through the first layer. Remember, however, that the random motion

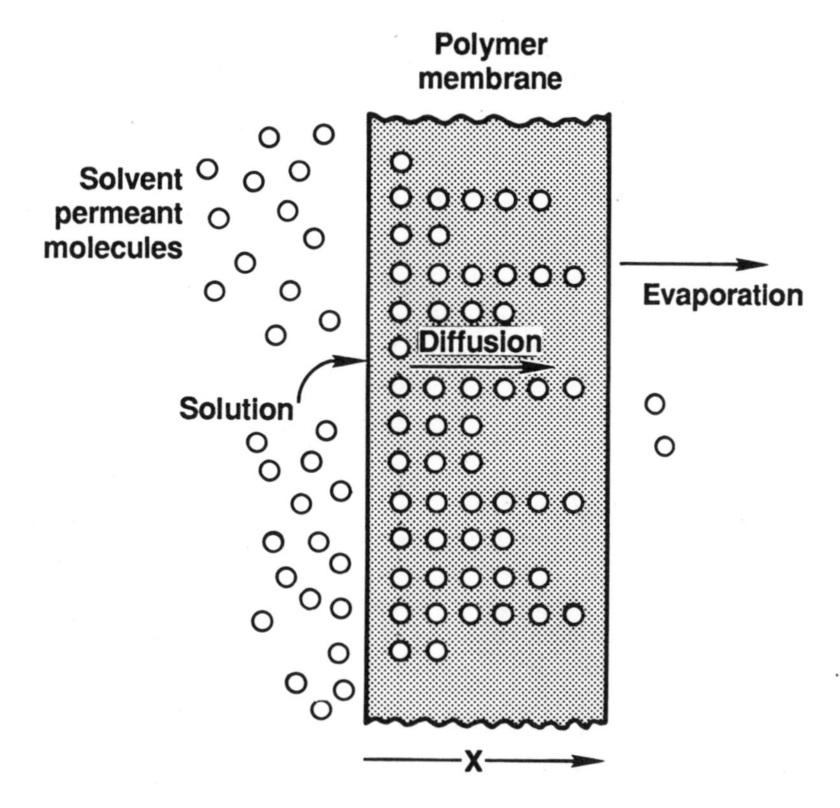

Figure 4-1. The permeation process.

of the solvent molecules may move them backward or forward. As shown in Figure 4-1, a concentration gradient usually develops. The rate of molecules evaporating from the downstream membrane surface in Step 3 is dependent on diffusion (Step 2) and solution (Step 1).[10]

Other terms are used to describe interaction of the solvent and polymer in CPC applications: *degradation*, *penetration*, and *permeation*. Degradation is also a function of the cohesive energies (solubilities) of the solvent and polymer. During degradation, the polymer may disintegrate because of solvent action on the polymer itself or on plasticizers or other additives in the polymer. Generally, the greater the degradation caused by a solvent, the more readily the solvent will permeate the polymer. Swelling usually takes place prior to degradation, and crosslinked polymers will swell but rarely degrade.

Penetration is a physical process by which bulk liquid solvent passes through the polymer membrane. This process is not dependent upon diffusion or solution, but instead relies on relatively large discontinuities that may occur in the polymer product because of construction processes. In effect, the solvent convectively flows through the pores with minimal interaction at the molecular level. Penetration is common in articles composed of paper, cotton, and other cellulose products, but can also occur through "pinholes," seams, and zippers in finished elastomers.

Permeation is the process by which a chemical passes through a polymer by means of solution and diffusion. In some cases, the solution step may be of little importance, as with N_2 passing through a thin sheet of polyethylene. However, for liquid solvents and CPC polymers, the importance of interaction or solution can vary greatly in magnitude. Since permeation is most dependent upon solution and diffusion, further detail about these processes is set forth in the following two sections.

The Solution Step

From solubility theory, two substances are soluble if, upon mixing, the free energy of the mixture is less than the sum of the free energies of the two pure substances. The free energy of mixing is defined as

$$\Delta G_m = \Delta H_m - T\Delta S_m, \tag{4.1}$$

where ΔG_m is the free energy of mixing, ΔH_m is the enthalpy, T is temperature, and ΔS_m is the entropy. The important point of this equation is that the enthalpy of mixing term (ΔH_m) must be less than the entropy term in order for the difference in free energy (ΔG_m) to be negative. The more negative the free energy difference, the better the solubility.

Solubility is dependent upon intermolecular forces. For example, if hydrogen bonds form between a polymer and solvent, the ΔH_m is small or negative, and solution (or negative ΔG_m) is assured. In effect, if these intermolecular forces are similar for two molecules, then solution may occur, since the enthalpy of mixing is quite small (or negative), causing the free energy to be negative.

The molecular forces that hold a liquid together are composed of three separate forces, as discussed in Chapter 3 (see Figure 3-2). The relationship of the three forces is

$$C^2 = D^2 + P^2 + H^2, \tag{4.2}$$

where C is the cohesive energy density, D is the dispersion parameter, P the polar parameter, and H the hydrogen bonding parameter.[11] The square root of the cohesive energy density is called the one-dimensional, or total, solubility parameter (SP). The three energy components taken as a vector are often referred to as the three-dimensional solubility parameter (3-DSP). Since permeation is a function of solubility and diffusivity (see Equation [4.9], below), it follows that by knowing something about the mutual solubility of a solvent and a polymer, one can determine something about its permeability.

An important feature of solubility parameter theory is that the total SP of a mixture is the weighted sum of the SP for the components of the mixture. Hence, if a paint product requires a solvent with a certain 3-DSP and no such solvent exists, a mixture of solvents may be made which closely matches that 3-DSP. By the same token, the polymers used in automotive products, such as carburetor O-rings, can be chosen such that they have the greatest amount of resistance to the chemicals with which they come in contact.

Simple Diffusion

There are many definitions for diffusion and many theories concerning the manner in which it takes place.[12] Put simply, diffusion is the random movement of molecules such that, given enough time, the distribution of molecules tends towards even concentration over space.

In contemplating the diffusion of molecules in a membrane, one must assume that the solution of the solvent and polymer occurs when a solvent molecule contacts the polymer surface. As the solvent diffuses through the polymer membrane, a concentration gradient is established that usually decreases linearly from highest concentration at the contacted surface to lowest concentration at the opposite surface. The reason for the concentration gradient is that the unexposed surface from which the solvent evaporates is being continuously swept by a carrier gas or liquid. If this did not happen, then no concentration gradient would occur, establishing instead a uniform and saturated distribution of solvent in the polymer. Of course the workplace situation may be somewhere between these two extremes, since under a protective garment there may be varying degrees of air movement.

According to some diffusion theories, a molecule in a "hole" within the polymer will move from that hole due to translational energy.[13] The molecule will move to another hole, provided that other holes in its vicinity are greater than a certain size. (Obviously, the molecule cannot occupy a hole that is smaller than itself.) The hole size formed by the polymer chains is a function of the energy of the chains. If the system is heated, the hole sizes will increase and the rate at which the holes are formed will also increase, aiding or speeding up the diffusion process.

The mathematical equation that describes simple diffusion is Fick's First Law,

$$J = -D\frac{dc}{dx} \quad , \tag{4.3}$$

where J is the mass flux in mg/min-cm^2, D is the diffusion coefficient in cm^2/min, c is the solvent concentration in the membrane in mg/cm^3, and x is the distance in centimeters from the contacted membrane surface. The minus sign indicates diffusion occurs toward areas of lower concentration. Consequently, the flux rate or diffusion rate is dependent upon the concentration gradient (dc/dx), and the diffusion coefficient (D) is simply a constant for the system. At constant temperature, the concentration gradient reduces to the concentration on the exposed surface divided by the thickness of the membrane (C/L), assuming that the opposite surface is continuously swept free of chemical (i.e., the concentration is 0 at that surface). For a liquid solvent contacting the membrane surface, the concentration at that surface is equivalent to the solubility of the solvent in the polymer. For a gaseous permeant exposed to the membrane surface, the maximum concentration in the membrane surface is the free concentration of the gas or vapor.

The diffusion coefficient is a constant that relates mass flux to concentration gradient and applies exclusively to the specific polymer and solvent in question. In other words, the diffusion coefficient derived for the permeation of benzene through neoprene will not be applicable to benzene through polyvinylchloride, or benzene through air. Diffusion coefficients are highly dependent on temperature.[10]

For most solvents that are liquids at room temperature, the diffusion coefficient can vary with the solvent concentration at the membrane surface. This is due to the chemical interaction of solvent and polymer. Consequently, the diffusion coefficient for a liquid in contact with a membrane will not be the same as for a non-saturated vapor of the same liquid in contact with the membrane. Most CPC permeation tests, however, use a liquid; this could be especially important in situations where the protective clothing is intended to prevent contact with a vapor.

If Equation (4.3) is integrated, the more familiar equation,

$$J = -D\frac{C_1 - C_2}{L} \tag{4.4}$$

is derived, where the diffusion coefficient D is not a function of concentration. Here, C_1 and C_2 are the permeant concentrations upstream and downstream, respectively, and L is the membrane thickness. As stated earlier, C_2 is usually 0 and falls out of the equation. In the cases described above, where D is a function of concentration,

$$\bar{D} = -\frac{1}{L}\int_{C_2}^{C_1} D\,dC\,. \tag{4.5}$$

In this equation, the diffusion coefficient in the integral is related to concentration by some other empirical equation. Consequently, in cases where the diffusion coefficient is dependent upon concentration, $\bar{D}$ is used to indicate an average diffusion coefficient.

D or $\bar{D}$ can be determined by measuring the concentration in the exposed surface of the membrane and the flux rate. In the strictest sense, ideal diffusion (D) will take place only if the permeant does not interact physically with the polymer or polymer additives, such that the concentration gradient is altered. Any non-ideal diffusion behavior is called non-Fickian diffusion and probably occurs where polymer swelling is noticeable.

Another procedure for measuring diffusion coefficients is the *lag time method*. As shown in Figure 4-2, lag time is the intersection point obtained by extending the steady-state permeation line (slope) to the x-axis. From diffusion theory, the following equation relates diffusion coefficients to lag time:[12]

$$D = \frac{L^2}{6T} \tag{4.6}$$

where L equals the thickness of the membrane and T equals the lag time.

Concentration in Equations (4.3), (4.4), or (4.5) is related to pressure by Henry's law,

$$C = Sp \tag{4.7}$$

where C is concentration, p is partial pressure, and S is the solubility coefficient. In the case of a solvent

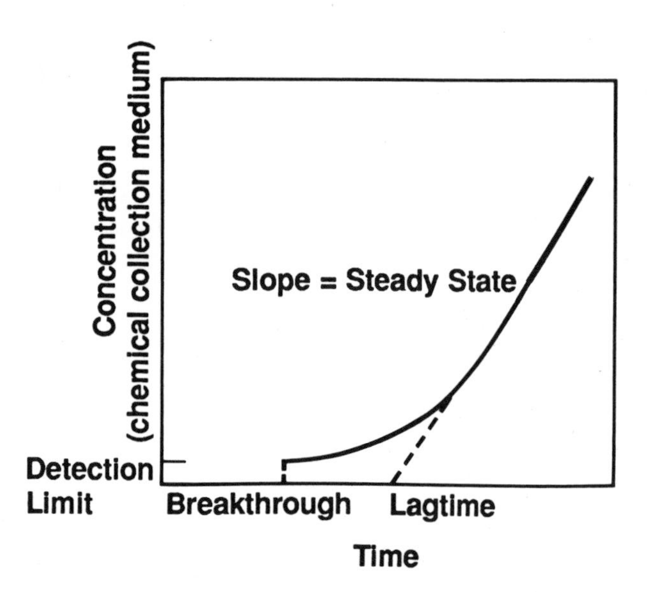

Figure 4-2. The lag time is the intersection point obtained by extending the steady-state permeation slope to the x-axis.

diffusing through a polymer that is continuously swept from the opposite side, C is the concentration of the solvent at the polymer surface, and p is its partial pressure. By substituting for C in Eq. (4.3), a new equation can be written:

$$J = DS\frac{dp}{dx} \qquad (4.8)$$

The coefficient DS is renamed P for permeability coefficient; thus

$$P = DS. \qquad (4.9)$$

The permeability coefficient is defined as a function of diffusion and solubility.

Molecular size and shape are known to affect diffusion rates. As might be expected from diffusion theory, the larger a molecule, the less likely the chance that a large enough hole will be formed to allow the molecule to diffuse. Consequently, diffusion coefficients are usually smaller as the molecule becomes larger in a homologous series.[14] Larger molecules also possess less kinetic energy, so that for a larger molecule the speed with which it is able to move to the next hole is less than that of a smaller molecule.

If an interaction occurs between polymer strands that are oriented effectively parallel to each other, then the possibility of hole formation of a certain size is reduced. This commonly occurs with polymers having a high degree of crystallinity. For example, high-density polyethylene has a greater crystallinity than low-density polyethylene. Plasticizers—which increase the space between polymer strands and make the polymer more pliable—are necessary to "soften" these polymers for use in protective clothing. While increased crystallinity usually decreases the diffusion coefficient, the addition of plasticizer increases the diffusion coefficient. Glassy polymers like PVC are below their glass-transition points at room temperature (see Chapter 3). Plasticizers must also be added to soften PVC for use in protective clothing.

Plasticizer considerations can be complex. For example, a chemical may have an initially high permeation rate, but at the same time it may interact with the plasticizer and leach it from the polymer. As more plasticizer is leached away, the permeation rate begins to decrease because the polymer strands become closer together. This is one example of how additives to polymers can cause non-Fickian diffusion, as mentioned above.[15]

Another polymer property that affects diffusion is crosslinking. Most accelerants used in crosslinking are organic, or crosslinks can occur directly between two polymer chains (covalent bonding). Crosslinking typically lengthens diffusion times because the polymer chains are pulled close together. However, if the crosslinking agent is a large organic molecule, the effect on diffusion can be more like that of a plasticizer. Also of importance is the idea that solvents can interact with the crosslinking agents. The net result may be non-Fickian diffusion.

The Evaporation Step

Permeation is often described as a repeating three-step process involving sorption of solvent molecules into holes forming between polymer chains, desorption of the solvent molecules, and then diffusion to the next available hole.[10] However, if the downward-gradient side of the polymer is constantly swept clear, the desorption or evaporation step likely plays a very small role in the rate determination of the overall permeability process.

Solubility is important both in terms of the solvent-polymer interaction during diffusion and the subsequent evaporation of the solvent from the opposite side of the polymer. A solvent or permeating substance with a very low vapor pressure will be slow to desorb from the opposite side. However, such a solvent will also be slow in diffusing and, because of possible poor molecular interaction with the polymer, may also be poor in dissolving into the polymer. Thus, the evaporation step would also appear to be a function of solubility and diffusivity.

The Importance of Solubility and Diffusivity

Previous attempts to predict how well a solvent will permeate through a polymer have relied on the concepts of solubility and diffusivity. However, the diffusion and solubility concepts have rarely been combined for the purpose of predicting permeation through chemical protective clothing.

One might expect that a chemical permeating through a membrane would be affected most by its size relative to the holes that are forming in the membrane. Since the size of any hole formed is presumably a random process with respect to some statistical mean, smaller molecules will have a greater selection of hole sizes through which they can diffuse, and consequently will diffuse faster. By the same argument, for a homologous series of molecules, larger molecules should diffuse more slowly (assuming that molecular interactions or solubilities are equal among the different species). For example, propane (a small molecule) permeates fastest, and n-pentane (a large molecule) permeates slowest for a propane to n-pentane series against polyisobutylene.[14] Since the molecular interactions between these solvents and the polyisobutylene are presumably very similar because of chemical structure, then the major factor affecting permeation is the diffusion coefficient.

Molecular shape is also important. A branch on a molecule may impede the diffusion process more so than the main body of the molecule. For example, permeation experiments using a homologous series of phthalic acid esters reveal that there is an optimum size for the length of the ester molecule.[16] In the case where C_4 is the ester molecule, diffusion is fastest, but for appending groups smaller or larger than C_4, diffusion becomes slower. This is presumably because the smaller appendages do not sufficiently shield the C=O group, thereby promoting molecular interaction, and the longer branches effectively slow the diffusion process and add nothing to the shielding.

Diffusion and solubility are independent of one another in the sense that a highly soluble solvent may not necessarily have a rapid diffusion rate. By the same token, a molecule with very little solubility in a polymer may have a very rapid diffusion rate. For example, small diatomic gases or noble gases diffusing through polymers often do so quite rapidly, even though the gases apparently have no interaction with the polymer. The size of the noble gas becomes important with respect to diffusion.

When molecules are quite small, such as methanol, carbon disulfide, and acetone, permeation rates are unusually rapid when compared with rates predicted by solubility alone.[7] The molecules may be small enough that the physical interaction (i.e., solution) with the polymer is insignificant in the overall permeation rate. In these cases diffusion is the most important step, resulting in a much higher permeation rate than for a larger molecule where chemical interaction or good solution is probably necessary for the molecule to gain entrance to the polymer.

Consequently, solubility and diffusion, though not directly related, appear to be very much interwoven in the permeation process, and neither can be considered alone. This fact may be one reason why some previous attempts to predict permeation have been only partially successful. Most of these attempts have relied singularly on either diffusion coefficients or solubility parameters, rather than addressing the joint action of the two concepts.

Other Variables in the Permeation Processes

Two other variables must be considered in the permeation process: the thickness of the polymer and the temperature of the solvent/polymer system. The effect of temperature on the permeation process can be related by an Arrhenius equation,

$$P = P_0^{-E_p/RT} \quad , \qquad (4.10)$$

where E_p is the activation energy of permeation, P is the permeation coefficient, P_0 is the pre-exponential factor, R is the ideal gas constant, and T is the absolute temperature. Similar equations may be written for the diffusion coefficient and solubility coefficient. When natural logarithmic transforms of temperature data are plotted against permeation coefficients, a straight line with intercept (P_0) and slope ($-E_p/RT$) is obtained.

Two examples in Tables 4-1 and 4-2 demonstrate this phenomenon. The temperatures shown in these tables are not uncharacteristic of those found in many manufacturing processes. Most laboratory permeation data are reported at 25°C; however, as can be seen in Table 4-2, a glove that was tested at 25°C in the laboratory may give vastly different results at 40°C. Note that the temperatures used here are test temperatures (solvent, polymer, and collection media.) Since many industrial processes are performed at higher than ambient temperatures, gloves coming in contact with solvents from those processes should be selected properly with respect to the actual temperatures of the solvents.

Thickness is also an important variable. The user of protective clothing should be aware that permeation rates and breakthrough times reported for a neoprene glove of a given thickness will not be applicable to a neoprene glove of another thickness. In the past, attempts have been made to standardize permeation data (such as the breakthrough time) by squaring it, or dividing it by the thickness of the glove or the square of the thickness of the glove. The latter relationship is derived from Eq. (4.6), which shows that the lag time (see Figure 4-2) is related to the square of the thickness of the membrane. This equation holds only for Fickian diffusion, so in many cases its use is incorrect, and more importantly, there is no relationship between breakthrough time and lag time.

The use of the thickness-squared term (or any correction term), if appropriate, should yield a

Table 4-1. Benzene permeation through neoprene. Data from Ref. 17.

Solvent T (°C)	Breakthrough Time (min)	Steady-State Flux (mg/cm^2-min)
7	40	0.19
22	24	0.23
37	16	0.33

Table 4-2. Methylene chloride permeation through Viton (unpublished data from author).

Solvent T (°C)	Breakthrough Time (min)	Steady-State Flux (mg/cm^2-min)
25	61	0.12
30	47	0.16
35	35	0.20
40	29	0.21

constant value when used for a series of breakthrough times for varying polymer thicknesses. Table 4-3 shows that this is not the case. The data in this table, derived from Ref. 15, show average breakthrough times and permeation rates for three thicknesses of neoprene latex gloves and three thicknesses of PVC gloves manufactured by Pioneer. For the neoprene gloves, 17 values of the breakthrough time and the permeation rate, derived for 17 different chemicals, were averaged over each of the three thicknesses. If division by thickness or thickness squared is an appropriate way to standardize for thickness differences, then the number derived from this quotient should be a constant over the range of thicknesses. As seen in Table 4-3, however, the number is not a constant. The same error can be seen for the PVC gloves, in which case only five sets of data (chemicals) were available for averaging.

Of the correction factors shown for permeation rate, PR × L is theoretically correct, if Fickian diffusion is occurring. This gives fairly good agreement for the neoprene gloves, but yields more variability for the PVC gloves. The relationship between thickness and permeation parameters is not constant, owing to experimental error and non-Fickian behavior. Consequently, one should be very careful when applying thickness correction factors. Ideally, permeability coefficients would be the best parameter for comparing gloves of varying thicknesses (see Eqs. [4.8] and [4.9]), but they are rarely reported.

Table 4-3. Standardization of permeation parameters for thickness.[a]

Polymer	Thickness (mm)	BL[b]	B/L	B/L^2	PR[b]	PR × L	PR × L^2
PVC	0.13	4.3	33	234	1199	156	20
	0.20	9.05	45	226	616	123	25
	0.31	16.76	54	174	399	124	28
Neoprene	0.29	7	50	84	146	42	12
	0.47	17.2	37	78	82	39	18
	0.70	28.4	40	60	64	45	31

[a] Data for averages of data from Nelson et al. for PVC (n=5) gloves 4, 5, 6 and Neoprene (n=17) gloves 15, 16,19.
[b] B = Breakthrough time in minutes; L = thickness, PR = Permeation rate in mg/min-cm².

Attempts to Model Protective Clothing Permeation

Colletta et al. attempted to derive short-term diffusion coefficients (values for D that occur prior to steady state—see Equation [4.6]) for nine carcinogens in order to predict the permeation resistance of protective clothing.[17] Unfortunately, they concluded that, due to the complexities associated with permeation through "filled" elastomers, accurate diffusion coefficients could not be determined, and modeling was impractical.

However, solubility tests (in which a polymer was immersed in a solvent for an extended period of time) did show interesting predictive results. The weight gain of the polymer was an indicator of the solubility of the solvent in the polymer. When weight gain and solubility were high, short breakthrough times were usually observed; however, the inverse was not so consistently true, and no correlation studies were performed.

Weeks and McLeod did correlation studies of weight gain and breakthrough times.[18] A typical result of their work is shown in Figure 4-3 for data involving Arochlor 1254 (PCB); they did not report correlation coefficients. Other chemicals studied were 1,1,2-trichloroethane, 1,2-dichloroethane, and 1,1,1-trichloroethane.

Holcombe studied the weight gain of Viton after its submersion in six different chemicals.[19]

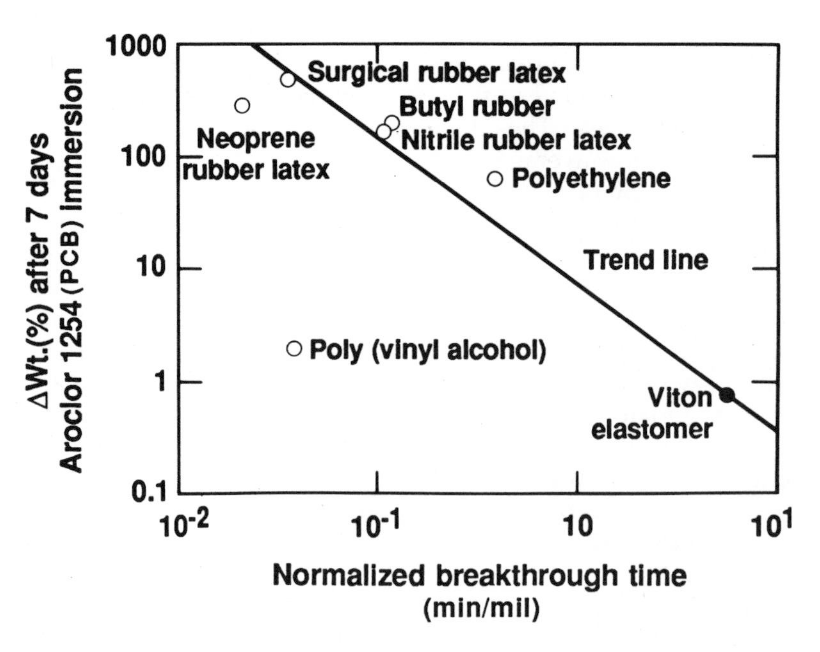

Figure 4-3. One correlation study of weight gain and breakthrough time.

When the natural logarithm of these values was correlated with breakthrough time, a correlation coefficient of 0.9 was obtained.

Recalling that "like dissolves like" from the discussion of the 3-DSP, one might conclude that if solvent and polymer have similar 3-DSP values, then the solvent should be highly soluble in the polymer. Henriksen used this concept in correlation studies with permeation parameters.[20] His preliminary work with permeation data from other researchers indicated that the natural logarithm of the steady-state permeation rate when correlated with polymer/solvent 3-DSP differences gave a correlation coefficient of (–)0.5 to approximately (–)0.7 The greater the difference between 3-DSP values for solvent and polymer, the lower the solubility, hence the negative correlation coefficients. Although these coefficients are not unity, they nevertheless indicate a dependence of permeation on solubility.

Bomberger et al. used permeability coefficients,[21] mainly derived from the work of Nelson et al.,[15] and correlated those against 3-DSP differences of solvent and polymer pairs. This work yielded correlation coefficients of (–)0.5 to (–)0.98. The researchers thought that correlation coefficients could be improved if more data for low permeation rates were available. These data are lacking because hours, or in many cases days, are needed before steady-state permeation is reached. A good example is Viton and carbon tetrachloride, where steady-state rate was reached after approximately seven days and breakthrough occurred after three days.[7] Henriksen's study also had the same problem concerning the lack of long breakthrough-time data.[20]

Perkins et al. also modeled permeation using the 3-DSP with the polymer Viton.[7] This work yielded a correlation coefficient of 0.69 for the ln values of breakthrough time, and a correlation of (–)0.65 for ln steady-state permeation rates when correlated against 3-DSP differences of polymer and solvent. In this work, solvents with long breakthrough times were included, without a great improvement in correlation coefficients.

In the studies of Perkins et al.,[7,22] Bomberger et al.,[21] and Henriksen,[20] solubility was determined by calculating the 3-DSP differences between the polymer and the solvent. If this difference is great, the solvent should not be very soluble in the polymer. The correlation coefficients obtained indicate the potential usefulness of the solubility parameter as a predictor of permeation. Indeed, Perkins et al. showed that by using confidence limits placed on the regression line, one can predict both the breakthrough time and the permeation rate. While the confidence limits were somewhat large, nevertheless a 95% lower confidence limit on breakthrough time, and a 95% upper confidence limit on permeation rate, could be used to predict the probable value of those two parameters for an untested solvent. For example, if 3-DSP values for the solvent and polymer alone are known, a minimum breakthrough time of 340 minutes (95% confidence), and a maximum steady-state permeation rate of 4.3 mg/m^2s can be predicted for xylene and Viton.

Perkins et al. also used polymer thickness and solvent molecular volume in combination with 3-DSP differences to predict permeation parameters.[22] Molecular volume is thought to affect diffusion rate. In these studies of butyl, Viton, and nitrile rubbers:

1. Thickness was not significant in breakthrough time studies, but was significant in predicting steady-state permeation rate;
2. Molecular volume was significant in breakthrough time studies, but it was only significant for nitrile in permeation rate studies;
3. 3-DSP difference was by far the most significant variable;
4. When significant variables alone are considered, coefficients of determination (R^2) were 0.38 to 0.55 for breakthrough time, and 0.57 to 0.64 for steady-state permeation rates.

All of the above results indicate that 3-DSP has utility, but only in semiquantitative applications. It may be used to predict which chemical will have the highest permeation rate for a polymer, or which polymer is most protective for a chemical.

In a good semiquantitative application of 3-DSP, Henriksen designed a new glove to protect against a chemical mixture.[20] In this case, the mixture consisted of various chemicals used in an epoxy resin process. Henriksen found that no single polymer protected against the hardener, the binder, or the various solvents used. Using 3-DSP data, he determined that a triple-laminate of polyethylene, polyvinyl alcohol–ethylene copolymer, and polyethylene should provide protection against all the chemicals of concern for an acceptable period of time. A single-use glove was fabricated of the laminate and was demonstrated to be acceptable.

One drawback in the 3-DSP approach and in most of the theoretical considerations discussed here is that glove products are not pure polymers. They contain plasticizers, antioxidants, fillers, pigments, and other additives. The solubility theory relates only the mutual solubility of the polymer and the solvent, and therefore acts as a predictor of permeation based on that two-component system. However, if a plasticizer is present in a polymer that has a substantially different solubility parameter, then results may not be predictable. If a solvent is selected that is poorly soluble in the polymer but readily soluble in the plasticizer, the solvent will leach out the plasticizer, changing the polymer structure. In the case of PVC, which contains large amounts of plasticizer (frequently 50–60%), this effect is prominent, resulting in a very stiff unplasticized PVC membrane, which cracks readily and is difficult to use.

This effect can be seen even for elastomers containing smaller amounts of additive. For example, the polymer Viton contains an amine-derivative additive; although Viton is resistant to many chemicals, it will degrade readily when exposed to certain amines such as diethylenetetramine. Solubility theory predicts that the amines would have little solubility in pure Viton; however, they readily decompose the glove polymer through their interaction with the amine.[19]

Once an additive is removed from a polymer, thereby changing the polymer configuration, the diffusion step is also affected, since the hole size and the hole-formation rate will also be changed. With some CPC polymers that contain plasticizer, a rapid breakthrough may occur due to the diffusion of the chemical through holes created by the plasticizer. If the plasticizer is then leached out by the solvent, the holes will close. Hence the steady-state permeation rate in this instance is very low, while the breakthrough time is short. The polymer film is also hardened and cracked by the deplasticizing action.[18]

Since polymer recipes vary among manufacturers, one must calculate 3-DSP values for all glove products of a generic polymer. At present, values for 13 polymers have been published and it is hoped that all manufacturers will publish data for their gloves.[23]

Another problem in modeling permeation is that neat or pure compounds rarely occur in the workplace. Workers are usually exposed to mixtures of compounds, and gloves or protective clothing must be selected for those mixtures. Solubility parameters, as noted above, can be derived for mixtures, and this approach is used successfully in the selection of solvents for paint formulations.

However, the permeation of chemicals through glove polymers may be different. A solubility parameter calculated for a mixture of two chemicals may not be a good predictor of permeation parameters. In other words, chemical A and chemical B may give permeation parameters that are predictable by their relative solubilities in the polymer, but the solubility parameter for the mixture of the two may not predict the total permeation of the mixture. More research is needed.

The few permeation studies involving mixtures have so far yielded interesting results.[15, 24, 25] Although these results may be summarized, it should be emphasized that permeation data involving mixtures are few, and any conclusions drawn from them are tenuous. Nevertheless, for a mixture containing two very similar chemicals with respect to solubility parameters (such as, for instance, toluene and xylene) the permeation is approximately additive. If the mixture contains a chemical that readily permeates the polymer and another compound that does not readily permeate the polymer, the results are not additive (see Figure 4-4). For example, if only small amounts of a good permeant are added to large amounts of a poor permeant, the poor permeant may begin permeating the polymer sooner than expected. In Figure 4-4, if permeation were additive, then one should obtain a line with zero slope for the sum of the permeation rates. In addition, a pure compound that does not break through in eight hours (permeation rate ~ 0) should not break through if the compound is mixed with another one.

However, this is not the case. A good permeant may be changing the structure of the polymer internally to the extent that the calculated solubility parameter for that polymer is no longer applicable. This process is similar to the manner by which a plasticizer affects a permeant, by opening up holes and allowing for greater diffusion. Such a permeant, once absorbed by the polymer, may also interact with the poor permeant. Under these changed conditions caused by the good permeant, the poor permeant now interacts with the "new" polymer structure and passes through the polymer. These interactions may not be quantitatively predictable with a 3-DSP approach. However, a better understanding of the process occurring with mixtures must be achieved and when more mixture data are generated, the 3-DSP approach can be modeled more thoroughly.

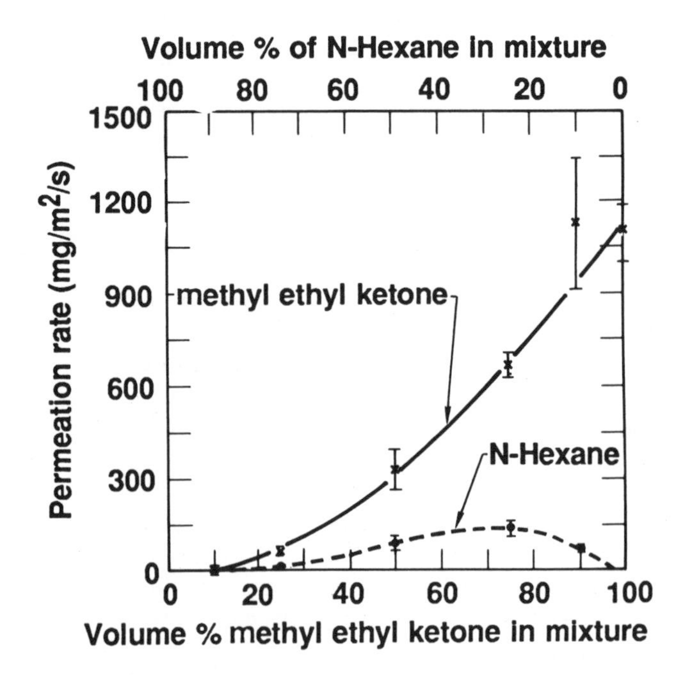

Figure 4-4. For each component in the mixture, steady-state permeation rate is plotted against mixture composition. As the MEK concentration in the binary mixture increases along the abscissa, the n-hexane concentration decreases.

Expectations for the Future

Since recipes for polymers are often trade secrets, at best only the *nature* of the additives and not the exact proportions may be known; this fact will continue to make modeling of permeation results difficult. One solution is to make the nature of polymer additives apparent with persistent research. Another solution is for manufacturers to be somewhat more lenient in allowing access to information, thus improving the ability of industrial hygienists to select protective clothing.

Industrial hygienists are trying harder to select the proper CPC to ensure workers' health and to reduce employers' liability. Consequently, vague guidelines for the selection of protective clothing are no longer useful. Industrial hygienists must have exact information which, for the most part, must come from manufacturers; therefore, manufacturers must be certain that the information they disclose is not misleading. Certainly, from the hygienists' point of view, manufacturers should willingly issue as much information about their protective clothing products as possible, even at the risk of losing some commercial advantage.

Industrial hygienists, however, must often deal with imprecision. Steady-state permeation rate and breakthrough-time data derived from permeation tests are inaccurate. In a review of permeation data, Bomberger et al. state that permeation parameters may vary as much as one order of magnitude when reported by different researchers for the same generic polymer-solvent combinations.[21] The many potential reasons for this discrepancy probably relate to differences in the permeation test cells used, glove lot differences, and detection limits. Permeation testing applied to protective clothing has begun only recently (see Chapter 5), and because of its infancy one can expect that the precision of test results should improve, along with our attempts to model those test results. In the meantime, users of this information must realize its imprecise nature and must be willing to err on the safe side when choosing protective clothing.

The user of permeation predictive tools must also realize that at present these can only predict the best polymer for a given application, rather than accurately quantify the permeation parameters. Therefore, the industrial hygienist should probably not select a particular piece of protective clothing because its breakthrough time is 60 minutes, for instance, and the job task will last only 50 minutes. Preferably, the industrial hygienist should select the best protective clothing for the application relative to other clothing types, and have knowledge of the permeation parameters, but at the same time the hygienist should realize that inaccuracy exists in those parameters, whether experimentally determined or predicted by a model.

With so many new developments in the areas of permeation testing and protective-clothing selection, it is not surprising that use of permeation test data is still at a very elementary level. Indeed, the entire personal protective clothing field is only in its infancy. One can look forward in the future to agreement in permeation test results among researchers, to increased knowledge of permeation theory among practicing industrial hygienists, and to improved methods for predicting permeation parameters. These factors should allow the industrial hygienist to make more rapid and more accurate decisions and should also remove the need for many permeation tests on the large number of chemicals and polymers now present in the workplace.

References

1. A.C. Stern, H.C. Wohlers, R.W. Boubel, and W.P. Lowry, *Fundamentals of Air Pollution*, Academic Press, New York (1973).

2. C.M. Hansen, "Solubility in the Coatings in Industry," *Farg Och Lach* **4**:69–77 (1971).

3. A. Beerbower and J.R. Dickey, "Advanced Methods for Predicting Elastomer/Fluids" in *Interactions*, ASLE Transactions **12**:1–20 (1969).

4. C.E. Rodgers, V. Stannett, and M. Szwarc, "The Sorption, Diffusion, and Permeation of Organic Vapors in Polyethylene," *J. of Poly. Sci.* **45**:61–82 (1960).

5. N. Cardarelli, *Controlled Release Pesticides Formulations*, CRC Press, Cleveland, OH (1976).

6. K.D. Reiszner and P.W. West, "The Collection and Determination of Sulfur Dioxide Incorporating Permeation and West-Gaeke Procedures," *Envir. Sci. Tech.* **7**:523–526 (1973).

7. J.L. Perkins, A.B. Holcombe, M.C. Ridge, M.K. Wang, and W.E. Nonidez, "Skin Protection, Viton, and Solubility Parameters," *Am. Ind. Hyg. Assoc. J.* **47**:803 (1986).

8. H. Burrell, "Solubility Parameter Values," in *Polymer Handbook*, J. Branderup and E. Immergut, Eds., John Wiley & Sons, New York (1975).

9. J. Hildebrand and R. Scott, *The Solubility of Non-Electrolytes*, 3rd ed., Reinhold, New York (1949).

10. H. Yasuda and V. Stannett, "Permeability Coefficients" in ***Polymer Handbook***, J. Branderup and E. Immergut, Eds., John Wiley & Sons, New York (1975).

11. C.M. Hansen, ***The Universality of the Solubility Parameter***, IEC Product Research and Development, **8**:2–11 (1968).

12. J. Crank and G.S. Park, "Methods of Measurement" in ***Diffusion in Polymers***, J. Crank and G.S. Park, Eds., Academic Press, New York (1968).

13. C.A. Kumins and T.K. Kwei, "Free Volume and Other Theories," in ***Diffusion in Polymers***, Academic Press, New York (1968).

14. S. Prager and F.A. Long, "Diffusion of Hydrocarbons in Polyisobutylene," *Am. Chem. Soc. J.* **73**:4072–4075 (1951).

15. G.O. Nelson, B.Y. Lum, G.J. Carlson, C.M. Wong, and J.S. Johnson, "Glove Permeation by Organic Solvents, *Am. Ind. Hyg. Assoc. J.* **42**:217 (1981).

16. K. Ueberreiter, "The Solution Process," in Diffusion in Polymers, J. Crank and G.S. Park, Eds., Academic Press, New York (1968).

17. G.C. Colletta, A.C. Schwope, I.J. Arons, J.W. King, and A. Sivok, *Development of Performance Criteria for Protective Clothing Used Against Carcinogenic Liquids*, US Department of Health, Education and Welfare, National Institute of Occupational Safety and Health, Pub. No. 79-106, Cincinnati, OH (1978).

18. R.W. Weeks and M.J. McLeod, *Permeation of Protective Garment Material by Liquid Halogenated Ethanes and a Polychlorinated Biphenyl*, US Department of Health, Education and Welfare, National Institute of Occupational Safety and Health, Pub. No. 81-110, Cincinnati, OH (1981).

19. A.B. Holcombe, *Use of Solubility Parameters to Predict Glove Polymer Permeation by Industrial Chemicals*, Masters Project, University of Alabama at Birmingham, School of Public Health (1983).

20. H.R. Henriksen, *Selection of Materials for Protective Gloves, Polymer Membranes to Protect Against Epoxy Products*, Danish National Labor Inspection Service, Lyngby, Copenhagen DK-2800 (1982).

21. D.C. Bomberger, S.K. Brauman, and R.T. Podoll, *Studies to Support PMN Review: Effectiveness of Protective Gloves*, US Environmental Protection Agency, Office of Pesticides and Toxic Substances, Technical Directive 68, Washington, DC (1984).

22. J.L. Perkins, M.C. Ridge, and W.E. Nonidez, "Predicting Permeation Properties of Butyl, Viton, and Nitrile Rubbers," presented at the 2nd International Symposium on the Performance of Protective Clothing, American Society for Testing and Materials Committee F23, Philadelphia, PA (1987).

23. J.L. Perkins and A. Tippitt, "Use of the Three-Dimensional Solubility Parameter to Predict Glove Permeation," *Am. Ind. Hyg. Assoc. J.* **46**:455 (1985).

24. R.L. Mickleson, M.M. Roder, and S.P. Berardinelli, "Permeation of Chemical Protective Clothing by Three Binary Solvent Mixtures," *Am. Ind. Hyg. Assoc. J.* **47**:236 (1986).

25. M.W. Spence, "Chemical Permeation Through Protective Clothing Material: An Evaluation of Several Critical Variables," presented at American Industrial Hygiene Conference, Portland, OR (1981).

Chapter 5

Test Methods

Norman W. Henry III

Introduction

Workplace protective clothing items, e.g., hoods, boots, or full ensembles, can be selected properly only when performance data indicate that resistance to chemicals lasts for the duration of anticipated, worst-case chemical exposures. This requirement for matching performance directly with possible exposures mandates a need for rigorously tested protective clothing.

Test methods, in helping to determine just what levels of protection are available, should mimic workplace conditions as closely as possible. Temperature and other appropriate environmental parameters (i.e., humidity) should be duplicated; actual chemicals of interest should be used in proper form and concentrations; and most importantly, exposure by the chemicals to protective clothing materials should be on the normal outside surface only.

One-sided exposure is an important point when testing chemical protective clothing. In actual use, spills, splashes and leaks will wet or coat an item's outside surface. The chemicals rarely, if ever, come in contact with edges or inside surfaces. (Actually, if edges or inside surfaces were to be contacted, the clothing has then failed in its mission of providing protection.) To simulate end-use conditions truly, test method designs should accommodate this fact and isolate all such inner areas. This chapter brings together contemporary ideas on testing protective clothing. It highlights a strategy for conducting tests and reviews the current status of efforts to develop test methods that are representative of workplace conditions.

A Brief History of Testing

In the protective clothing business, both manufacturers and marketing people proclaimed the advantages of their newly developed clothing materials for use in the workplace environment. Formerly, however, their claims for protection were often not substantiated by test methods that demonstrated performance of a particular material against hazardous chemical substances.

Traditionally, both protective clothing manufacturers and users have depended on simple immersion tests as primary sources for performance information on chemical resistance. Because such tests permit edge and inside-surface contact, data has been generated and published that can be inaccurate and misleading. Immersion tests are particularly inappropriate for evaluating multi-layered composites.

Furthermore, protective clothing manufacturers often marketed their products on the basis of the results of crude test methods designed to measure physical hazard resistance. Consequently, protective clothing items would be selected on their ability to withstand abrasion, cut, and puncture. There might or might not have been an attempt to match physical strength with the results of immersion testing.

Much of the information developed from immersion and physical-property tests exists in currently available literature; because of this, many occupational health and safety professionals are still selecting protective clothing items that do not represent an optimum match between exposures and materials.

A number of manufacturers and a few users developed their own test methods and data reporting schemes based on convenient criteria for simple apparatus design and testing conditions for in-house applications and, perhaps, external marketing efforts. Often, these criteria bore little resemblance to users' actual work environments. Minimum emphasis was placed on standardization, which makes comparison of data from one source to another difficult, if not impossible. This also made the selection process confusing for users, especially for those without laboratory facilities of their own. Also, the identification of performance goals for new product development had been, at best, inexact for the manufacturers.

One of the earliest publications describing a comprehensive test method for evaluating the performance of protective clothing was reported by Adrian L. Linch.[1] He described a permeation test method for measuring the breakthrough time and diffusion rate (permeability) of protective clothing materials exposed to hazardous chemicals. This was the first time that a chemical permeation resistance was recognized as an important criterion for measuring the degree of protection that clothing can provide. It was also the first reference to a standard permeability test cup procedure for determining chemical resistance (Figure 5-1). Results of some of Linch's studies led to the development of a "Chem-Proof" air suit used for complete isolation from the work environment. The lack of routine or comparable test methods for the evaluation of protective clothing has made the selection process imprecise. Some surveys have indicated that one out of every four workers in the United States is exposed to some form of skin irritant, and that one percent of these workers develop skin disorders from such exposures. Furthermore, occupational diseases accounted for two-thirds of all job-related diseases, and seven out of ten industrial claims paid by insurance companies were for temporary disability resulting from dermatitis.[2,3]

Some Action

Liability issues, as well as the swirl of confusion surrounding measurement of protective clothing performance, has prompted an aggressive movement toward standardizing test methodology and reporting. Since the 1970s, manufacturers, users, and governmental agencies have been working together to address broad issues relating to protective clothing performance. In the United States, the principal forum for these activities is the American Society for Testing and Materials (ASTM) Committee F23 on Protective Clothing. In Europe, the comparable group is the International Standards Organization; in Scandinavian Countries, it is the Nordic Coordinating Group on Protective Clothing as a Technical Prophylactic Measure (NOKOBETEF).

United States governmental organizations, such as the National Institute of Occupational Safety and Health (NIOSH) sponsored a program to investigate the performance of protective materials against selected carcinogenic chemicals. This program originated as a result of the Occupational Safety and Health Administration

Figure 5-1. (a) Some sample test cells.

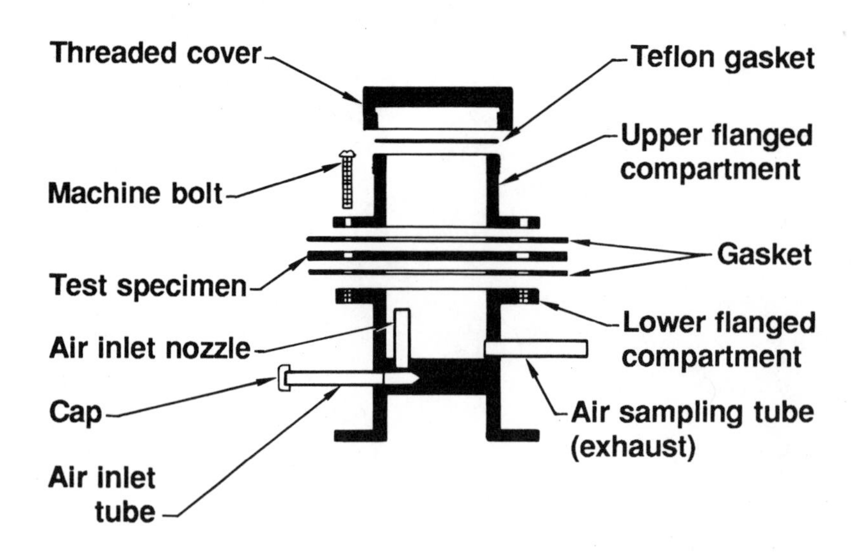

Figure 5-1. (b) Schematic of the permeability test cup.

(OSHA) recommendation that protective clothing materials be "impermeable" to carcinogenic liquids. Since OSHA did not provide guidelines for assessing the barrier properties of protective clothing materials, NIOSH sponsored a study on the development of performance criteria for protective clothing used against carcinogenic liquids. A technical report on this subject was written and published in October 1978 by NIOSH.[4] Results of this work include the following:

1. A description of the form and extent of worker needs for protective clothing.
2. A recommended criterion for clothing resistance to permeation and an overview of several other functional requirements.
3. Recommended test methods for evaluating the performance of clothing materials.
4. Data showing the resistance of currently used clothing materials to permeation by selected carcinogens.

In 1986 NIOSH listed dermatitis as one of the ten leading work-related diseases in the United States and sponsored a national symposium to plan a strategy for the prevention of work-related dermatological conditions. Other governmental organizations are now showing interest in the standardization of protective clothing. These include the Environmental Protection Agency, the US Coast Guard, the Federal Emergency Management Agency, and the Department of Transportation. Their efforts have been directed to participation in ASTM's F23 committee. Private groups such as the American National Standards Institute and the National Fire Protection Association have also expressed interest and provided some support for the concept of standardized test methods for protective clothing. Interest in evaluating protective clothing has also been apparent in articles published by leading occupational health and safety journals.[5–12]

A Testing Strategy

At this time in the test method development process, each candidate material for use in protective clothing must be tested against each chemical to which it will be exposed. This is a far-reaching statement, because the possible combinations of materials and chemicals are limitless. Later developments may well identify shortcuts and techniques for accurately predicting performance within material categories or within chemical groupings. However, at the present time, such useful correlations are mostly speculation. Therefore, testing is now being driven by a one-on-one strategy in four parts:

- Begin testing with flat specimens of candidate materials. Specimens should consist of either a single layer or, when appropriate, a composite of multiple layers arranged in the proper order.
- Apply the challenge chemical on only the outside of the candidate materials. In actual use, spills, splashes, and leaks will wet or coat an item's outside surface; but the chemical rarely, if ever, comes in contact with edges or inside surfaces. To truly simulate the correct end-use conditions, test method designs should accommodate this fact and isolate all such areas.
- Take only the best candidate materials through an entire testing protocol, presuming that such a protocol will involve testing both chemical resistance and physical-hazard resistance. Complex, expensive test methods should be reserved for only the best performers.
- Upgrade from testing flat material specimens to testing assembled end-items as the final confirmation of design and construction integrity.

Following this strategy, current testing of chemical protective clothing is based on a hierachy of three parameters: resistance to degradation, resistance to penetration, and resistance to permeation. Testing by these three performance parameters presents candidate clothing materials with challenges of increasing sophistication and rigor. Only the best performers are expected to undergo permeation testing, since this method is the most expensive and time-consuming laboratory protocol. Not only does this approach represent an organized plan for evaluating performance of both candidate materials and end items, but it also provides a route for exercising overall cost containment during what can be a lengthy testing process.

Existing Test Methods

To date, test-method development efforts have been oriented toward benchtop, laboratory techniques. With the availability of sophisticated analytical equipment, this focus has permitted an examination of many variables involved in mimicking workplace conditions. A second generation of methods, still some time away, will use these first laboratory methods as a springboard into field testing and lower costs. For now, however, chemical resistance and physical hazard resistance are being evaluated in the laboratory. See Tables 5-1 and 5-2 for a complete list of current physical and chemical test methods.

Table 5-1. Physical property test methods.

Property	Test method (ASTM #)
Abrasion resistance[a]	D3389
Blocking	D1893
Brittleness	D2137
Bursting strength	03786
Coating adhesion	D751
Cut resistance[a]	F23.20.01[b]
Durometer	D2240
Flammability[a]	D568
Flex fatigue[a]	D671 (Plastics)
	D430 (Elastomers)
Hydrostatic resistance	D751
Low temperature bending	D2136
Ozone resistance	D3041
Penetration resistance	F903
Puncture propagation tear	D2582
Puncture resistance[a]	F23.20.02[b]
Seam strength[a]	D751
Stiffness[a]	D1043 (Plastics)
	D1053 (Elastomers)
Tear strength[a]	D751
Tensile strength[a]	D751 (Supported materials)
	D412 (Unsupported materials)
Thickness	D751
UV light resistance	G26
Weight	D751
Zipper strength[a]	2061

[a] Key physical property.
[b] Test method is currently being developed.

Table 5-2. Test methods for chemical protective clothing.

Characteristics	Test
A. Chemical resistance	
1. Permeation resistance	ASTM F739-85: Resistance of Protective Clothing Materials to Permeation by Hazardous Liquids and Gases
2. Swelling and solubility	ASTM D471-79: Rubber Property—Effects of Liquids
3. Strength degradation	ASTM D543: Resistance of Plastics to Chemical Reagents
4. Crazing	ASTM F484-77: Stress Crazing of Acrylic Plastics in Contact with Liquid or Semi-Liquid Compounds
5. Transparency	ASTM 1746-70: Transparency of Protective Clothing Materials to Penetration by Liquids
6. Penetration resistance	ASTM F903-87: Resistance of Protective Clothing Materials to Penetration by Liquids
B. Strength	
1. Tear resistance and strength	ASTM D751-73: Testing of Coated Fabrics
	ASTM D412-75: Rubber Properties in Tension
Fed. 191A-5102	(ASTM D1682): Strength and Elongation, Breaking of Woven Cloth: Cut Strip Method
Fed. 191A-5134	(ASTM D2261): Tearing Strength of Woven Fabrics by the Tongue Method
2. Puncture resistance	See Ref. 4
3. Abrasion resistance	ASTM D1175: Abrasion Resistance of Textile Fabric
C. Dexterity/flexibility	
1. Dexterity (gloves only)	See Refs. 4, 26, 27
2. Flexibility	ASTM D1388: Stiffness of Fabrics, Cantilever Test Method
D. Aging resistance	
1. Ozone resistance	ASTM D3041-72: Coated Fabrics—Ozone Cracking in a Chamber ASTM D1149-64: Rubber Deterioration—Dynamic Ozone Cracking in a Chamber
2. UV resistance	ASTM G27: Operating Xenon-Arc Type Apparatus for Light Exposure of Non-Metallic Materials—Method A—Continuous Exposure to Light
E. Chemical Selection Guide	ASTM F1001-86
F. Whole ensemble	Standard Practice for Pressure Testing of Gas Tight Totally Encapsulating Chemical Protective Suits
1. Pressure (Inflation) testing	ASTM F1052-87
2. Qualitative leak testing[a]	ASTM F23.50-02
3. Quantitative leak testing[a]	ASTM F23.50-03
4. Qualitative evaluation of fit, function and integrity	ASTM F23.50-04

[a] Test Method is currently being developed.

Chemical Resistance

Most workers involved in the production, use, and transportation of chemicals are exposed to numerous compounds capable of causing harm on contact with the human body. The effects of these chemicals can range from acute trauma, such as dermatitis or burn, to chronic degenerative diseases, such as cancer or pulmonary fibrosis. Since engineering controls will not eliminate all exposures, attention must be given to minimizing the risk of direct skin contact through the use of tested protective clothing that resists degradation, penetration, and permeation.

Degradation

Degradation is the resistance to deterioration of the physical properties of clothing materials upon contact by a chemical. Significant properties that can be evaluated include thickness, weight, elongation, tear strength, cut resistance, and puncture resistance.

Two highly qualitative test methods have formerly been used for evaluating resistance to degradation. The most frequently used, the immersion test discussed earlier in this chapter, is the traditional technique for testing rubber and plastic for chemical resistance, ASTM D543.[13] For gloves, an "inverted glove" method has been used, which requires a large volume of test chemical. These two tests can result in visible evidence of deterioration. Materials may get harder, stiffer, and more brittle, or they may soften, weaken, and swell to several times their original dimensions. These visible changes provide rough, qualitative evidence of degradation.

However, a new semi-quantitative test method for degradation is now under evaluation by ASTM's Committee F23. With this more quantitative method, the resistance of a protective clothing material to degradation by a liquid chemical is determined by (1) measuring the thickness, weight, and elongation of specimens of the clothing material; (2) contacting additional, separate specimens of the material with the chemical of interest; and (3) measuring the thickness, weight, and elongation of the additional specimens to identify changes resulting from contact with the chemical. The list of physical properties can be expanded to include resistance to abrasion, cut, puncture, and the like. The method will be used to provide an evaluation of flat specimens cut from candidate CPC materials, or cut from finished items of protective clothing. Materials that demonstrate acceptable resistance should then be evaluated further to determine resistance to penetration and permeation.

Penetration

Penetration is the resistance to flow of a chemical on a *non-molecular* level through closures, seams, pinholes, or other imperfections in a protective clothing material. The way in which protective materials are glued, seamed, stitched and heat-sealed are all important to the overall performance of the end item. Penetration is an area often overlooked, but ASTM has now promulgated Standard Test Method F903-87.[14] The resistance of protective clothing material to penetration by a liquid chemical is determined by subjecting one side of a material specimen to the chemical under a pressure of 1 psig and noting the time at which visible penetration occurs. In the test apparatus (Figure 5-2), the material specimen acts as a partition separating the liquid chemical from the viewing surface. After charging the hazardous liquid chemical into the test cell, pressure is applied, and the time at which a drop of the liquid is seen on the inside surface is recorded. Adding a coloring agent such as a dye before running the test aids in discerning the drop of liquid chemical. The test determines resistance to penetration using a simple pass/fail criterion. Penetration, or failure, indicates flow through seams, pinholes, or other imperfections. Whole end-item penetration tests involving spraying of a hazardous liquid chemical have been investigated but are not considered practical from a safety standpoint, since a hazardous chemical is involved and a suitable exposure chamber or hood would be required. A solvent "splash" method that uses 100 µl has been reported, but it has not been fully evaluated.[15]

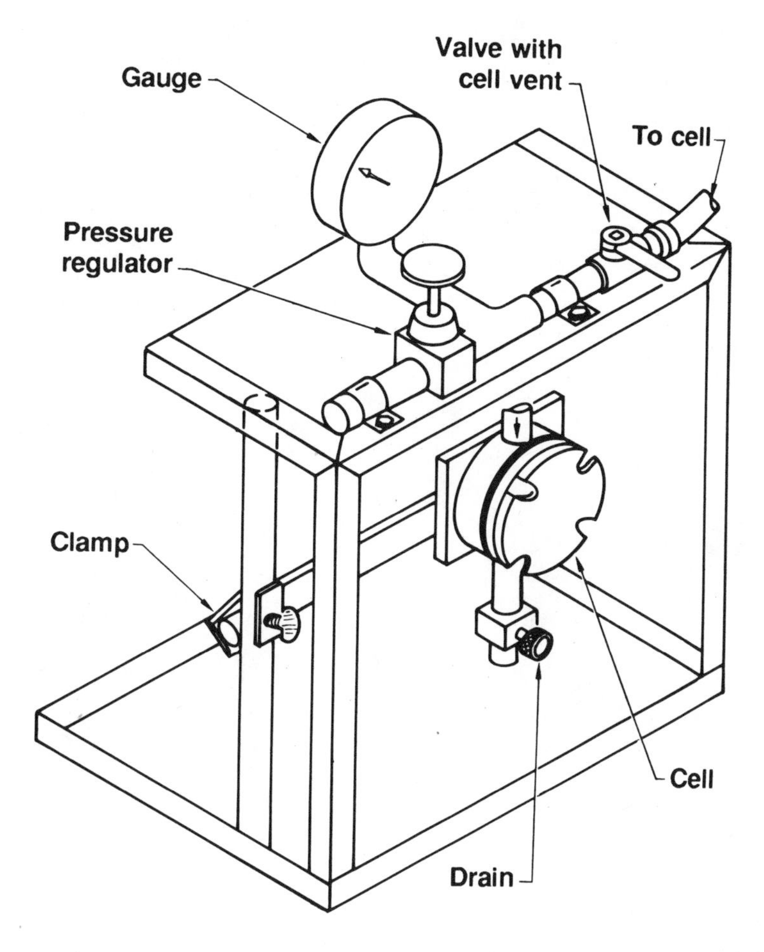

Figure 5-2. (a) Detailed drawing of the ASTM penetration test apparatus.

Permeation

Permeation is the resistance to chemical movement through a protective clothing material on a *molecular* level via absorption, diffusion, and then desorption. Permeation is best explained mathematically by Fick's law of diffusion, which states that "the rate of mass diffusion through a unit surface area of clothing material (or membrane) is proportional to the concentration gradient of the liquid chemical (permeant) across the material."[16] The mathematical expression is as follows:

$$J = -D\frac{dc}{dx}$$

where J is the mass flux in mg/cm^2,
D is the diffusion coefficient in cm^2/s,
c is the liquid chemical concentration in the clothing material in mg/ml, and
x is the distance in cm from the contacted membrane surface.

The minus sign in the equation accounts for a decreasing c as x increases. In other words, as the thickness of the material increases, the amount of liquid chemical able to permeate decreases. Diffusion coefficients are known for many polymeric materials and can be found in polymer handbooks.[17] However, values for many chemicals used in the workplace are not known. Designated as ASTM Standard Test Method F739, a permeation test method was the first and most complex procedure developed by Committee F23.[18] With its most recent revision, this method can be used to evaluate a material challenged by either liquids or gases.

Figure 5-2. (b) Photograph of the ASTM penetration test apparatus.

The resistance of a protective clothing material to low-level permeation can be determined by measuring the initial breakthrough time of the chemical and then monitoring the rate of subsequent permeation through the material specimen. In the test apparatus, the material specimen acts as a barrier separating the chemical of interest from a medium that collects the chemical as it permeates. The collecting medium, which can be either a liquid or a gas, is analyzed quantitatively for its concentration of challenge chemical.

This analysis allows direct calculation of the amount of chemical that has permeated the material as a function of time after initial contact. The level of detectability of the chemical depends on the sensitivity of the analytical equipment used.

This test procedure requires that a material specimen cut from the protective clothing be clamped into a test cell (Figure 5-3) as a barrier membrane; the experimenter then contacts the "outside" of the specimen with a hazardous chemical, and tests the medium on the "inside" at intervals for the presence of permeated chemical. Depending on the collection medium chosen, suitable analytical techniques involving wet chemical, spectrophotometric, infrared, and gas or liquid chromatography are used to measure the amount of liquid chemical permeating the test specimen. The temperature of the test can also be controlled by immersing the test cell in a water bath maintained at actual use temperature. Temperature-related effects on permeation have been reported by other researchers.[19]

By measuring breakthrough time, the method can be used to estimate the duration of maximum protection provided by a protective clothing material under continuous chemical contact. Then, by measuring the permeation rate,

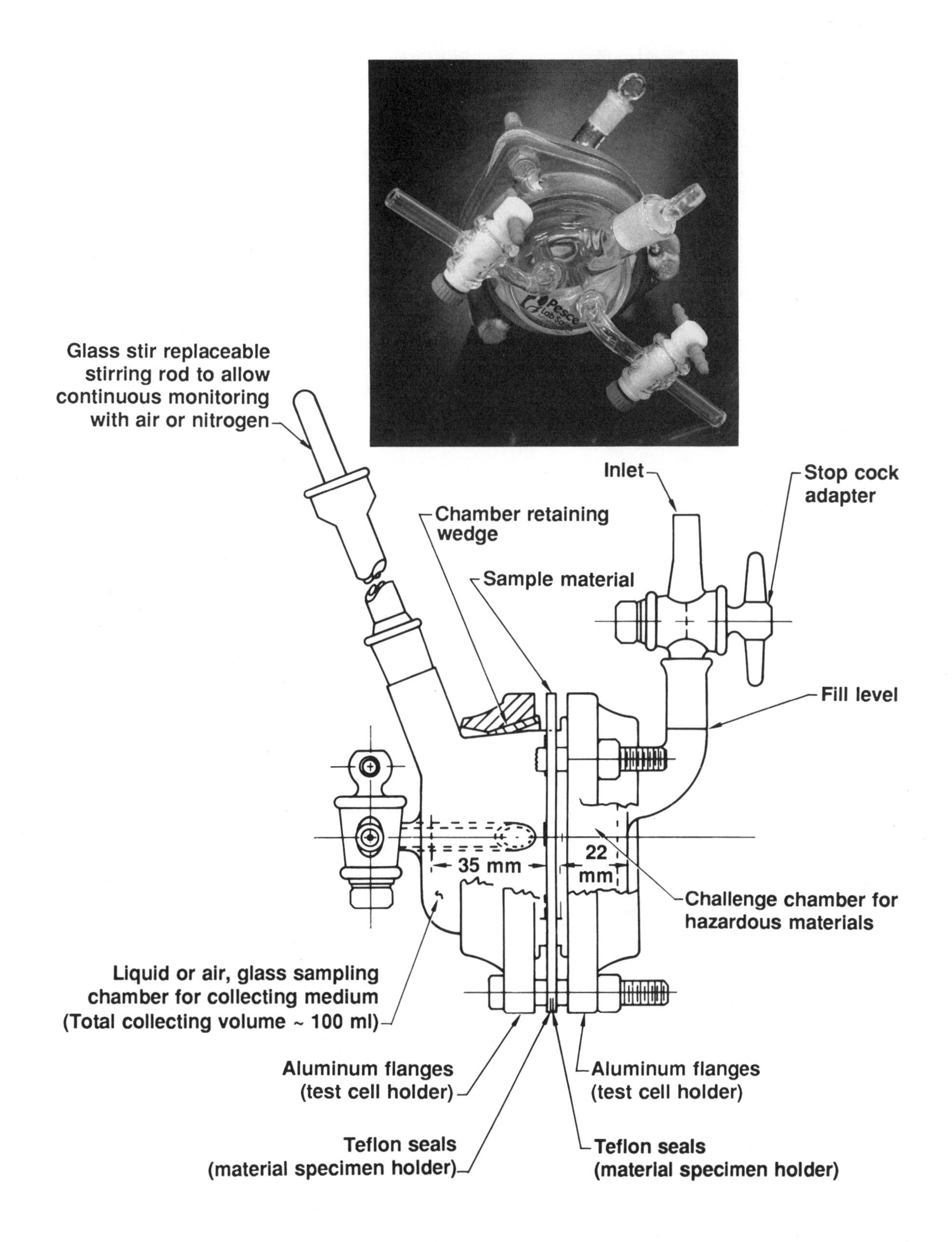

Figure 5-3. ASTM permeation test cell, schematic drawing and photograph.

the method can be used to identify protective clothing materials that limit potential exposures to acceptable, steady-state dermal contact levels. (NOTE: At present, such contact data are not available for most chemicals. Recognized occupational health authorities must derive the contact data from threshold limit values, 50% lethal doses, or other similar indices.)

On a somewhat larger scale, a whole-glove test method has been reported and shown to be effective in determining permeation resistance.[20] To measure the integrity of totally-encapsulating chemical protective suits, ASTM Committee F23.5 on Total Ensemble Testing has developed ASTM F1052, "Standard Practice for Pressure Testing of Gas Tight Totally Encapsulating Chemical Protective Suits." By inflating the suit to a prescribed pressure and determining the pressure drop over time, a measure of the suit's gas tightness can be obtained. This standard practice allows the suit user to carry out a pressure leak-rate test in the field before use or after cleaning and decontamination.

Physical Properties

The other main area of concern about protective clothing performance is the physical properties of the clothing material. The physical hazard resistance can be evaluated by familiar abrasion, puncture, cut, tear, and tensile test methods.[21] These tests measure the degree of strength that a material has while subjected to potential sharp, rough, and heavy objects that may be encountered on the job. Results of such tests give additional information about a material's performance capabilities, if a particular job involves handling various sharp and heavy objects.

Another issue involves testing the utility or functionality of clothing materials. Obviously, those materials without sufficient flexibility cause problems. A stiff, thick glove material may prevent exposure, but the material may be so thick that the wearer may not be able to turn a switch or pick up a tool.

Furthermore, comfort of the clothing material should be considered. Protective clothing that is not breathable or comfortable causes worker stress. This stress can be reduced by evaluating materials for their breathability at various climatic conditions. At present, no generalized stress-test method has been developed. Perhaps methods will be forthcoming from the newly formed ASTM Committee F23.51 on Human Factors, which is addressing such issues as comfort, sizing, and dexterity.

However, many test methods for evaluating a wide range of physical properties of materials have been developed by numerous private and public groups, though most are not applicable to testing protective clothing because they do not take a proper accounting of workplace conditions. Accordingly, several other methods that correct this deficiency have been proposed. The furthest along in the development process are those for measuring resistance to cut and resistance to puncture:

Cut. This draft test method is under review by Committee F23. A thorough scrutiny of the document's provisions may result in the identification of an existing method that, with minor alteration, can be adapted to protective clothing. By the proposed method, the resistance of protective clothing material to cut is determined by measuring the force required to cause a sharp-edged blade to cut the surface of a material specimen. The document defined the blade configuration and the material specimen size, conditioning, and position on the test apparatus. The blade is to be made of tool-hardened steel with dimensions chosen to correspond to a common workplace cut hazard, such as the edge of a piece of broken glass.

Puncture. This draft test method is also under review by Committee F23. An existing test method might be adaptable here as well. By the current proposal, the resistance of clothing materials to puncture is determined by measuring the force required to cause a pointed, uniformly moving penetrometer to puncture a material specimen. The document defines the penetrometer configuration and the material specimen size, conditioning, and position in the test apparatus. The penetrometer is to be made of tool-hardened steel with dimensions selected to correspond to a common workplace puncture hazard, such as a 4d nail.

As noted previously in the discussion on degradation testing, methods for measuring performance such as cut and puncture can be integrated within the degradation protocol to evaluate changes that result from contact with chemicals.

Decontamination

Though considerable interest has been shown by groups such as firefighters and hazardous-waste response teams that are often required to re-use protective clothing, no formal test method has been identified for decontamination. A method should be developed for this purpose, but first the difficult problems associated with defining contamination and decontamination must be resolved. A general discussion of the decontamination of chemical protective clothing is presented in Chapter 8.

Automation

Automated test methods for permeation testing are a new development arising from the need to reduce the cost of testing and also to reduce the risk of exposure to operators when testing highly toxic materials. Since triplicate analysis is specified in all test methods to determine reproducibility, the need for automation quickly became apparent to those laboratories that intended to provide such testing as a service. With automated analytical equipment and computer systems available, data can be rapidly generated, analyzed, and stored. Such an automated chemical permeation system and test service is already being offered by the Radian Corporation in Austin, Texas. Another company that offers a fully automated permeation system is ProTech. A picture of their system (VOA200) is shown in Figure 5-4.

New Developments

In addition to the draft test methods for determining resistance to degradation, cut, and puncture, several proposals have been made for expanding the current permeation method. These have evolved from ASTM's Committee F23:

• The test method should be broadened to permit measurement of vapor and particulate permeation. Modifications would parallel the techniques in F739, but would allow for a vapor or a particulate challenge on the outside surface of the clothing materials. Modifications in the test protocol would permit continuous particle exposure on the challenge side and the development of a sampling system for particles on the inside collection medium. Various particle detection techniques are available for developing this procedure, but they need to be investigated for practical consideration as laboratory tests. A similar strategy for vapor testing is also being evaluated.

Several new test methods were reported at the Second International Symposium on Protective Clothing in Tampa, Florida in 1987 involving methods for low volatility and low solubility compounds, as well as a method using ^{14}C tracers. A method to measure the aerosol penetration of a porous fabric was also presented at this symposium.

• Alternative test cells have also been suggested for the liquid permeation test method.[22] Small test cells that permit the use of smaller volumes of hazardous chemical are safer

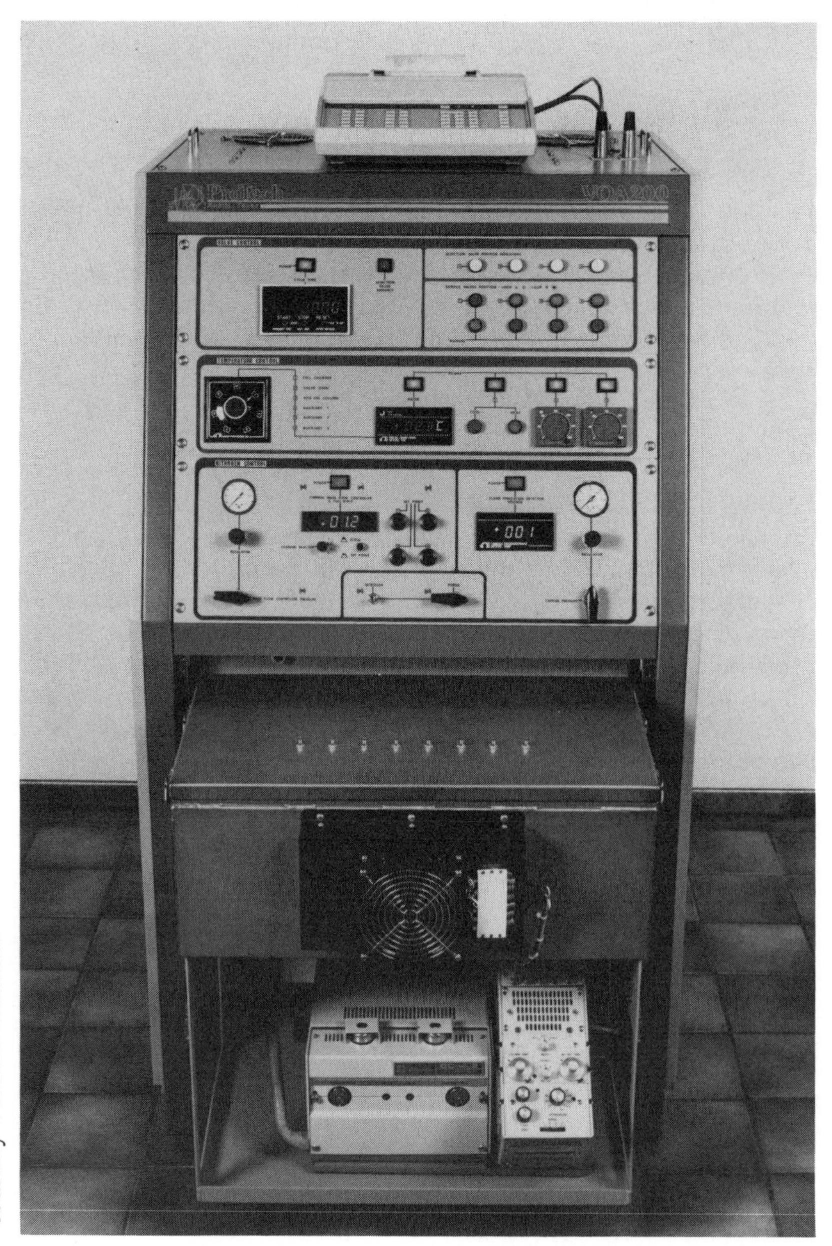

courtesy Protech Scientific

Figure 5-4. A fully automated permeation test system from ProTech (VOA200). (Photo courtesy Bruce Sorenson, Protech Scientific)

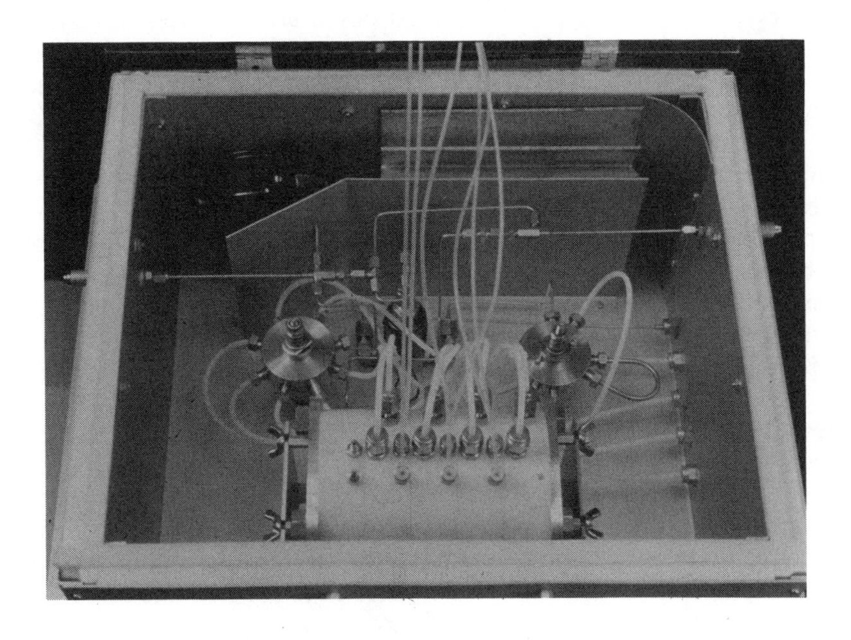

Figure 5-5. Microcell made by the Radian Corporation.

and more practical to use for multiple analysis. One proposed test cell modification requires only one milliliter of chemical to conduct a liquid permeation test (Figure 5-5). Materials used for the construction of the test cell are also important. The F739 test cell is currently suitable for most organic permeation studies because glass is not affected by most organic solvents. However, the glass cell would not be appropriate for hydrofluoric acid permeation studies, since this acid attacks glass. Stainless-steel cells would be acceptable and have been used specifically for such tests.

Eventually these alternate test cells need to be evaluated for equivalency to the original glass ASTM cell reported in method F739. Once equivalency has been demonstrated, the modified cells may be used. An equivalency test protocol is being considered by ASTM Committee F23.

Additional work by one investigator has shown that solubility parameters of materials are a critical variable in predicting performance.[23] Interest in this area can only help improve the protection properties of clothing materials, thereby reducing exposure in the workplace.

Further developments are already on the drawing board. The assimilation of chemical permeation resistance data into a master data bank has been suggested, a monograph listing permeation test data has been printed,[24] and a field guide for the selection of protective clothing has been published.[25] Such test data will be also be useful in filling out Material Safety Data Sheets (MSDSs). A pocket-sized, easy-to-use selection guide for chemical protective clothing has recently been published by Foresberg and Mansdorf. These documents provide an excellent opportunity for specifying chemical-specific barrier materials for protective clothing.

The development of new products has also shown growth recently as a result of the test methods.[27] New, more resistant protective clothing materials are being sought by investigators using the test methods as tools to evaluate performance.[28] To help in these evaluations, ASTM has developed a standard guide for selection of test chemicals, ASTM F1001-86. A list of the test chemicals is shown in Table 5-3.

One of the latest areas of test method development is in biological resistance of protective clothing. Workers in the health care profession are potentially exposed to a number of biological liquids, such as blood, which are capable of transmitting various microorganisms that may produce infections and disease, such as hepatitis or Human Immunodeficiency Virus (HIV). Gloves, surgical gowns, uniforms, and emergency response clothing are susceptible to penetration of blood-borne pathogens. Efforts to measure performance have focused on specific manufacturers' test methods, such as a strike-through test, and on users' methods developed by health care workers themselves. Currently ASTM subcommittee F23.70 on Biological Hazards is considering test methods for determining biological resistance. To date, these methods have focused on determining liquid penetration resistance by subjecting the protective clothing materials to a biological liquid or simulant at a prescribed pressure and time interval until visible droplets are observed. Materials showing no visible penetration are considered to have passed the test. This method and others are still under consideration by the ASTM subcommittee. It is possible that more than one test method will be needed for the evaluation of biological resistance because of the diversity of microorganisms and routes of transmission (i.e., aerosols).

Finally, as more testing is done and better methods are developed, greater protection will be afforded to those in the workplace. Once the degree of protection has been determined under controlled conditions in the laboratory, rather than accidentally in the workplace, health and safety professionals will have a greater sense of confidence in selecting protective clothing materials. Field monitoring methods are still needed to substantiate these laboratory test methods.

Table 5-3. ASTM F1001-86 standard battery of test chemicals.

1. Acetone
2. Acetonitrile
3. Carbon disulfide
4. Dichloromethane
5. Diethylamine
6. Dimethylforamide
7. Ethyl acetate
8. n-Hexane
9. Methanol
10. Nitrobenzene
11. Sodium hydroxide (50%)
12. Sulfuric acid (fuming)
13. Tetrachloroethylene
14. Tetrahydrofuran
15. Toluene

Acknowledgment

The author wishes to acknowledge the suggestions and comments of Gerard C. Coletta.

References

1. Adrian L. Linch, "Protective Clothing,"*CRC Handbook of Laboratory Safety* (CRC Press, Boca Raton, FL, 1971), pp. 124–137.

2. Julian B. Olishifski, *Fundamentals of Industrial Hygiene,* 2nd ed., National Safety Council, Washington, DC (1971), p. 67.

3. Julian B. Olishifski, *Fundamentals of Industrial Hygiene,* 2nd ed., National Safety Council, Washington, DC (1971), p. 202.

4. Gerard C. Coletta, Arthur D. Schwope, Irving J. Arons, Jerry W. King, and Andrew Sivak, "Development of Performance Criteria for Protective Clothing Used Against Carcinogenic Liquids," U.S. Department of Health, Education, and Welfare, Washington, DC, *DHEW (NIOSH) Publication No. 79-106* (1978).

5. John J. Croley, "Protective Clothing Responsibilities of the Industrial Hygienist," *Am. Ind. Hyg. Assoc. J.* **27**, pp. 140–143.

6. Robert W. Weeks, "Permeation of Methanolic Aromatic Amine Solutions Through Commercially Available Glove Materials," *Am. Ind. Hyg. Assoc. J.,* December 1977, pp. 721–725.

7. Eric B. Sansone and Y.B. Tewari, "Permeability of Laboratory Gloves to Selected Solvents," *Am. Ind. Hyg. Assoc. J.*, February 1978, pp. 169–174.

8. Wayne B. Bosserman, "How to Test Chemical Resistance of Protective Clothing," *National Safety News*, September 1979, pp. 51–55.

9. John R. Williams, "Permeation of Glove Materials by Physiologically Harmful Chemicals," *Am. Ind. Hyg. Assoc. J.* **40**, pp. 877–882.

10. John R. Williams, "Chemical Permeation of Protective Clothing," *Am. Ind. Hyg. Assoc. J.*, December 1980, pp. 884–88.

11. Norman W. Henry III and Nelson C. Schlatter, "Development of a Standard Method for Evaluating Chemical Protective Clothing to Permeation of Hazardous Liquids," *Am. Ind. Hyg. Assoc. J.* **42**, pp. 202–207.

12. Gary O. Nelson, B.Y. Lum, G.J. Carlson, C.M. Wong, and J.S. Johnson, "Glove Permeation by Organic Solvents,"*Am. Ind. Hyg. Assoc. J.* **42**, pp. 217–225.

13. American Society for Testing and Materials, "Resistance of Plastics to Chemical Reagents, " ASTM Method D543, Part 35, American Society for Testing and Materials, Philadelphia, PA (1987).

14. American Society for Testing and Materials, "New Standard Test Method For Resistance of Protective Clothing Materials to Penetration by Liquids," ASTM Method F903, American Society for Testing and Materials, Philadelphia, PA (1987).

15. Eric B. Sansone, "Resistance of Protective Clothing Materials to Permeation by Solvent 'Splash,'" *Environmental Research,* **26** (1981), pp. 340–346.

16. John Crank and G.S. Park, *Diffusion in Polymers* (Academic Press, New York, 1968) pp. 1–39.

17. Allen F. M. Barton, *CRC Handbook of Solubility Parameters and Other Cohesion Parameters,* (CRC Press, Boca Raton, FL, 1983).

18. American Society for Testing and Materials, "Standard Test Method for Resistance of Protective Clothing Materials to Permeation by Hazardous Liquid Chemicals," ASTM Method 739, Part 46, American Society for Testing and Materials, Philadelphia, PA (1985).

19. B. Alexy and R.M. Buchan, "Temperature-Related Permeability of 1,1,1-Trichloroethane Through Chemical Protective Clothing," paper presented at the American Industrial Hygiene Conference, Cincinatti, OH, June 11, 1982.

20. John R. Williams, "Evaluation of Intact Gloves and Boots For Chemical Permeation," *Am. Ind. Hyg. Assoc. J.* **42,** pp. 468–471.

21. Gerard C. Coletta, Arthur D. Schwope, Irving J. Arons, Jerry W. King, and Andrew Sivak, "Development of Performance Criteria for Protective Clothing Used Against Carcinogenic Liquids, Summary of ASTM Methods" Washington, DC, US Department of Health, Education, and Welfare, *DHEW (NIOSH) Publication No. 79-106,* 1978, pp. 8–13.

22. Stephen P. Berardinelli, R.L. Mickelson, and M.M. Roder, "Chemical Protective Clothing: A Comparison of Chemical Permeation Test Cells and Direct-Reading Instruments," *Am. Ind. Hyg. Assoc. J.*, December 1983, pp. 886–889.

23. Mark W. Spence, "Chemical Permeation through Protective Clothing Material: An Evaluation of Several Critical Variables," paper presented at the American Industrial Hygiene Conference, Portland, OR, May 28, 1981.

24. W. Stephen Dixon, Norman W. Henry III, and Robert W. Pell, *Chemical Resistance of Protective Clothing,* Haskell Laboratory for Toxicology and Industrial Medicine, E.I. duPont de Nemours and Company, Wilmington, DE (1984).

25. Arthur D. Schwope, *Guidelines for the Selection of Chemical Protective Clothing, Volume 1, Field Guide,* American Conference of Governmental Industrial Hygienists, Arthur D. Little, Inc., Cincinnatti, OH, 1983.

26. K. Foresberg and S.Z. Mansdorf, *Quick Selection Guide to Chemical Protective Clothing,* (Van Nostrand Reinhold, New York, 1989).

27. Krister Foresberg, "Development of Safety Gloves for Printers," ERGOLAB Report 511:10, Stockholm/Gateborg, Sweden, November 1981.

28. L. Sterling, "Test Program for Work Gloves," Department of Occupational Safety, Division for Occupational Medicine, Labor Physiology Unit in Umea, Sweden, Research Report 1980:18, 1980 (Transcribed from Swedish).

Chapter 6

Chemical Protective Clothing Types, Methods of Construction, and Typical Applications

James S. Johnson and John Varos

Introduction

Chemical protective clothing (CPC) has become an important means of protecting the worker from chemical exposure. To be effective at providing the broad range of chemical protection needed in industry, current CPC technology has become quite sophisticated. Early attitudes seemed to be that if chemical protective clothing "isn't thick, black, and ugly, it isn't a good product." An item of CPC is no longer judged only by its color, thickness, and appearance. As part of the design of CPC, actual physical and chemical performance properties of candidate materials are now evaluated against hazardous chemicals.

Chemical protective clothing is now available in a broad range of sizes. CPC manufacturers have recognized the entry of women into the workplace and are now manufacturing equipment that is specifically designed to fit them. The typical CPC user is also growing increasingly knowledgeable, which forces the marketplace to respond with a broader range of products. CPC use by government entities such as the US Army, the National Aeronautics and Space Administration (NASA), and the US Coast Guard has also increased the available protective clothing materials and products.

Various types of chemical protective clothing are available to protect specific parts of the body that may be exposed to a chemical threat. An effective way to illustrate the major types of CPC in use is to start with a totally-encapsulating chemical protective (TECP) suit and identify the major components, as shown in Fig. 6-1. Table 6-1 lists the specific body parts that require protection and the type of CPC equipment used.

To provide effective chemical protection to the worker, one must carefully examine the performance of each CPC type being used. The following sections in this chapter will provide a more detailed description of each type of CPC equipment.

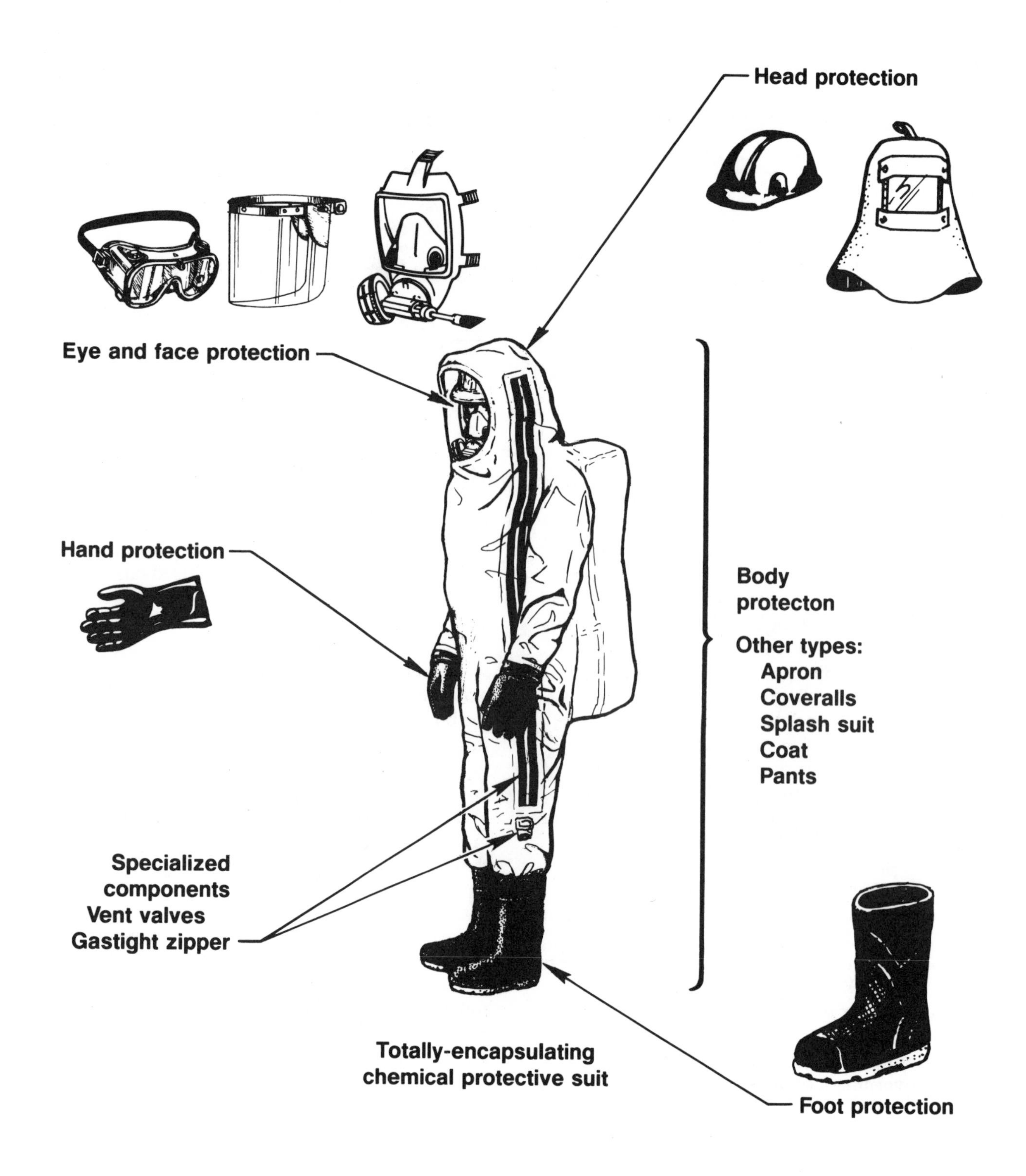

Figure 6-1. Types of chemical protective clothing.

Table 6-1. The specific body part(s) requiring protection and the type of CPC equipment used.

Body Part	CPC Type
Eye and Face	Chemical goggle
	Faceshield
	TECP suit lens
	Full-face respirator
Head	Splash hood
	Hard hat
	TECP suit material and seams
Hand	Glove
Body	Apron
	Coveralls
	Splash suit (pants and coat)
	TECP suit material and seams
	TECP suit vent valves
	TECP suit gas-tight zipper
Foot	Boots/booties
	TECP suit material and seams

Eye and Face Protection

Many chemicals in the workplace can cause significant eye damage—including permanent loss of vision—if they are splashed into the eye. Facial scarring and disfigurement can also result from direct contact with these chemicals.

The chemical goggle is the workhorse of the CPC equipment used for eye protection. Goggles consist of a soft-sided lens holder designed to fit snugly to the face, providing a seal that prevents chemical exposure in the event of a splash. The lens holder is usually made from a soft pliable polymer, such as PVC, using injection-molding techniques. Most lens holders include air vents to allow moisture to escape. Some commercially available models incorporate splash guards over the air vents to prevent liquid entry into the goggle. Other commercially available models are gas-tight, and provide no venting of the enclosed eye area.

The faceshield is another common form of chemical eye and face protection used in industry. Unfortunately, most users do not understand that this type of CPC is designed to prevent only direct splash exposures to the face, and not to provide complete eye protection. Chemical safety glasses or goggles must also be worn under the faceshield. For instance, a splash, stream, or jet of chemicals could hit the worker's chest, splash up under the faceshield, directly into the wearer's eyes. Two lengths and several thicknesses of faceshields are commercially available. The longer and thicker lens provides better protection to the wearer.

The lens material must provide clear undistorted vision as well as chemical resistance. Initially, the clear vision requirement was the major consideration in selecting lens material; however, with recent increased concern about chemical hazards, the chemical resistance requirement is now equal to or, in some cases, more important than the clear vision properties.

During the design stage of their CPC ensemble for propellant-handlers, NASA recognized the importance of the chemical resistance of the visors. A series of hypergolic propellant exposure tests were carried out on candidate visor materials.[1] Figures 6-2 and 6-3 are representative photographs showing the effects of chemical exposure to the candidate lens materials. From these tests NASA was able to choose the best lens material for their needs. This illustrates that performance information from actual testing is necessary when highly reactive and highly toxic chemicals are being handled.

A recent incident at Benicia, California, involving the failure of TECP suit lenses illustrates the potential life-threatening consequences of a lens failure.[2] A leaking tank car of anhydrous dimethylamine required HazMat team response for evaluation and control. During the response, one of the participants noticed the cracking of his TECP suit lens. The four individuals involved in the response immediately exited the vapor cloud. One individual's lens cracked completely and exposed the interior of his TECP suit to the anhydrous dimethylamine vapor. The self-contained breathing apparatus (SCBA) under the suit protected the man's respiratory tract, but his unprotected skin was exposed for a short time to the corrosive vapor, resulting in a moderate case of dermatitis.

From an analysis after this incident, materials scientists noted that polycarbonate is known to craze and crack when exposed to amines. In Figs. 6-4 and 6-5, the actual damage to the TECP suit

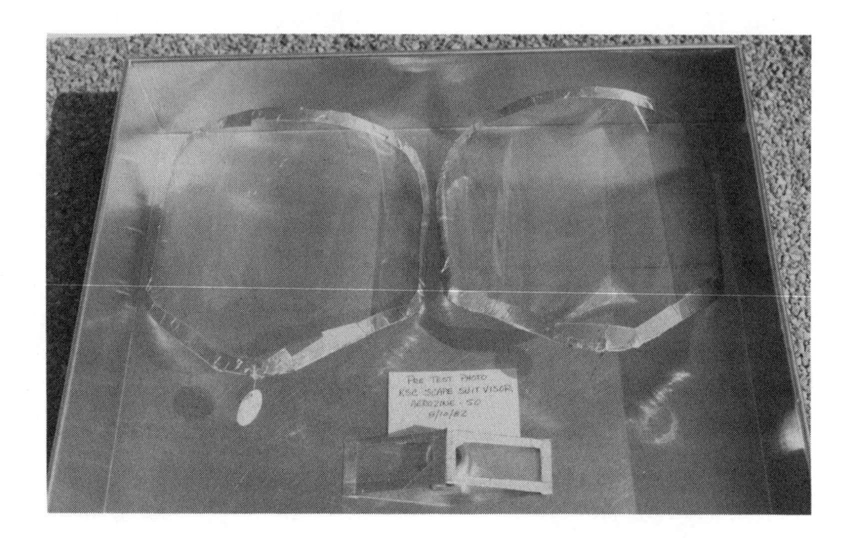

Figure 6-2. Unexposed NASA visor candidate material before exposure to a hypergolic propellant.

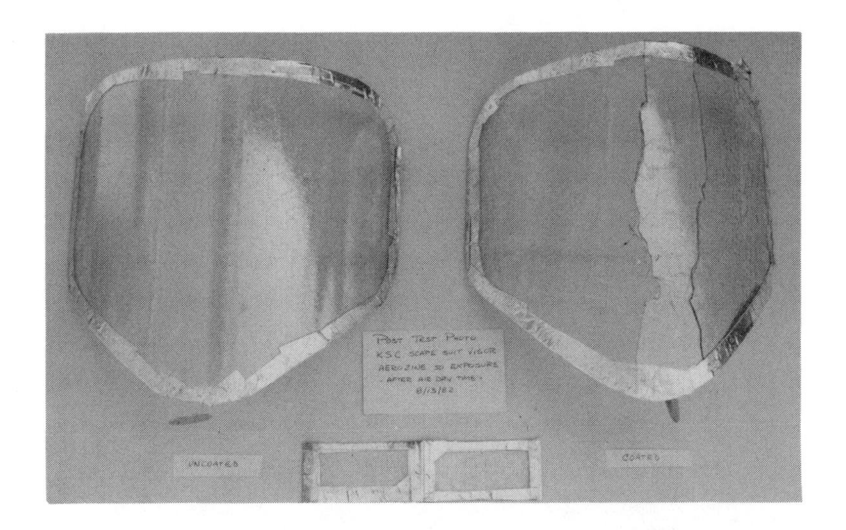

Figure 6-3. Exposed NASA visor candidate material after contact with a hypergolic propellant.

lenses and SCBA facepiece lenses can be seen. The SCBA facepiece lenses were attached to TECP suits in such a manner as to expose them directly to the anhydrous dimethylamine. This incident illustrates the importance of knowing the chemical resistance of lens material, especially in high-risk chemical responses.

Recent advances in goggle lens design incorporate double-lens construction with a vacuum in between to reduce fogging. Lens material made from a tetrafluoroethylene blend is used when a high degree of chemical resistance is required. The revised National Fire Protection Association Standard for SCBAs has included a test for facepiece lens scratch resistance.[3] This new requirement may also cause the development of lens materials that are more chemically resistant.

For a more detailed description of the polymers used in chemical eye and face protective equipment, see Chapter 3.

Figure 6-4. TECP suit lens crazing and cracking caused by exposure to anhydrous dimethylamine at Benicia, California, 8/12/83.

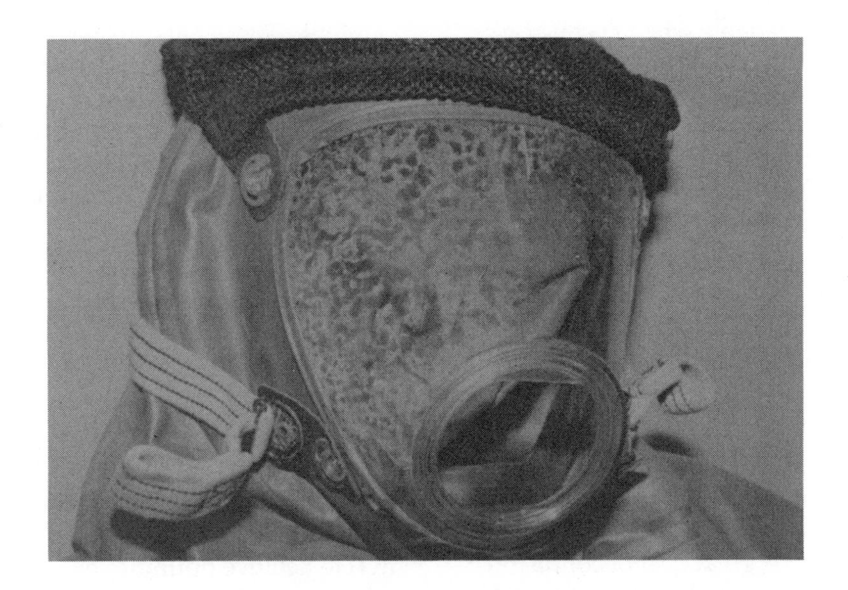

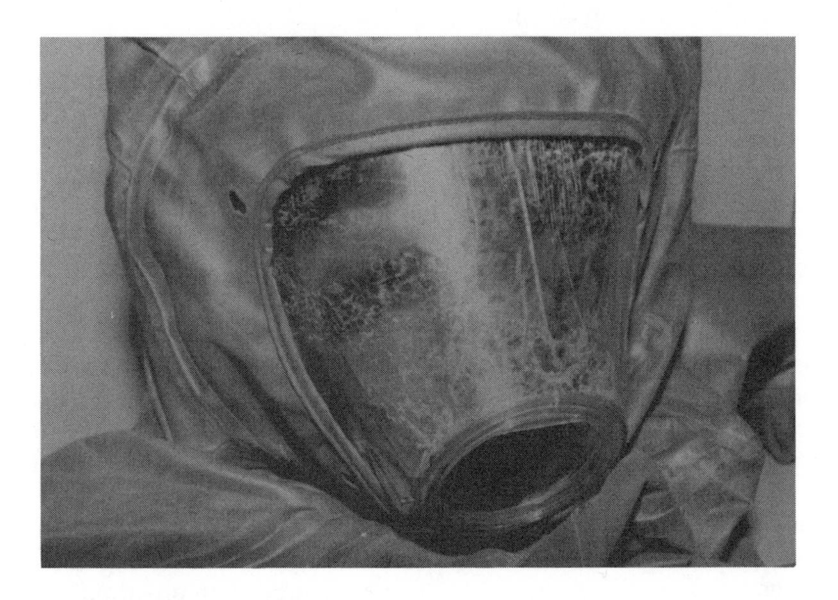

Figure 6-5. TECP suit SCBA facepiece lens crazing and cracking caused by exposure to anhydrous dimethylamine at Benicia, California, 8/12/83.

Head Protection

Chemical splash protection for only the head is not a common workplace requirement, but a splash hood is available when necessary. The splash hood is made from chemical-resistant material with a viewing lens incorporated in the front of the hood. The hood drapes over the shoulders and uses straps to hold it in place. The splash hood is normally used with a one-piece or two-piece whole-body garment to provide complete body protection from splash. Splash-protection clothing offers little, if any, gas or vapor protection, however.

Some manufacturers incorporate hard hats or bump hats into their chemical protective ensembles. These hats are generally made from polycarbonate and protect the head from bumps and falling objects, although they are not designed to provide chemical protection. The need to use these types of hats in conjunction with various forms of CPC should be evaluated on a case-by-case basis.

Hand Protection

Gloves are the most common form of CPC. A large variety of products is available from many manufacturers. A great deal of confusion exists as to what glove should be selected for a specific job. Before one can select the correct glove, however, the basic differences in chemical protective gloves must first be understood. Table 6-2 lists some of the more common chemical-resistant construction materials of chemical protective gloves.

For a given thickness, the polymer material selected has the greatest influence on the level of chemical protection provided by the glove. For a given polymer, a greater thickness will increase the level of chemical protection, if the subsequent loss in dexterity (caused by the thicker glove) can be safely tolerated for the job application. Additives to the raw polymer can be used during manufacturing to enhance the desired characteristics of the material. Because of this, some variation in chemical resistance and physical performance will occur in like polymer gloves from one manufacturer to another.

Mickelsen and Hall determined the breakthrough times of two common glove polymers using three chemicals. The largest difference they found was a factor of ten, 300 vs 30 minutes, for nitrile products tested with perchloroethylene.[4] Also, the precise level of temperature and processing time must be maintained during manufacturing to properly crosslink or cure (or fuse, for vinyl) to achieve optimum characteristics. This may result in some slight variations in performance from batch to batch of the same type of glove.

Chemical protective glove construction can be further classified into unsupported (including flock-lined) and supported or coated gloves. Impregnated gloves, a third type commonly used to protect against cuts and abrasions, are not considered liquid-proof, and are therefore not chemical resistant.

Unsupported gloves are made of a cured polymer film. These gloves offer a higher degree of dexterity and come in a greater selection of polymers, lengths, and thicknesses than supported gloves. For flock-lined gloves, the inside surface of the glove is coated with adhesive and covered with shredded processed cotton or synthetic fibers to improve extended wearing comfort and to facilitate donning and doffing. The thickness gauge of an unsupported glove is generally expressed as the single-wall thickness to the nearest thousandth of an inch, measured in the work area of the glove, avoiding any embossed pattern or raised grip surface where possible. Test method Federal Standard #191, Method 5030.2, is normally used.[5] Some typical CPC glove thicknesses and classifications are listed in Table 6-3.

Table 6-2. Common construction materials of chemical protective gloves.

Butyl,Isobutylene-Isoprene Rubber (IIR)
Natural Rubber (NR)
Neoprene, Chloroprene (CR)
Nitrile,Acrylonitrile-Butadiene Rubber (NBR)
Polyethylene (PE)
Polyvinyl Alcohol (PVA)
Polyvinyl Chloride (PVC)
Polyurethane (PU)
Viton[a]
Silver Shield[b]

[a] Du Pont Trademark.
[b] North Trademark.

Table 6-3. Typical chemical protective glove thicknesses and classifications.

Glove thicknesses (inches)	Classifications
<0.008	Ultra or very light weight
0.008—0.012	Light weight
0.012—0.018	Medium weight
>0.018	Heavy weight

Supported gloves or coated gloves are fabric liners with cured polymer coatings. Supported gloves can offer increased thermal, abrasion, and tear resistance at the sacrifice of dexterity. Supported gloves are manufactured in various styles using several polymers. Some of the more common styles are the gauntlet cuff, the knit wrist cuff, and the safety cuff, as shown in Fig. 6-6.

The liner construction is generally either knit (interlock or jersey) fabric or woven fabric of cotton, wool, or synthetic blends (such as polyester or rayon). The fabric weight is described in ounces per square yard. The more common styles of liner construction are listed below.

Two piece—The glove liner is made from two pieces of material sewn in the general form of a glove.

Clute cut—The front of the glove liner is made from one piece while the back is sewn from a number of parts. This style provides a formed glove and usually has a knit wrist.

Gunn cut—The back of the glove liner is made from one piece. The second and third fingers are set in with a seam across the bottom of each. The seams in the finger areas extend two-thirds of the way around each finger, eliminating exposed seams in the wear areas.

The length of a chemical protective glove is another important consideration. The length is measured from the tip of the middle finger to the cuff edge of the glove. Some typical glove lengths and the protection they offer are listed in Table 6-4.

Table 6-4. Typical glove lengths and the protection they offer.

Protection	Length
Hand protection only	up to 12"
Midarm protection	13–15"
Elbow length protection	16–18"
Shoulder length protection	30–32"

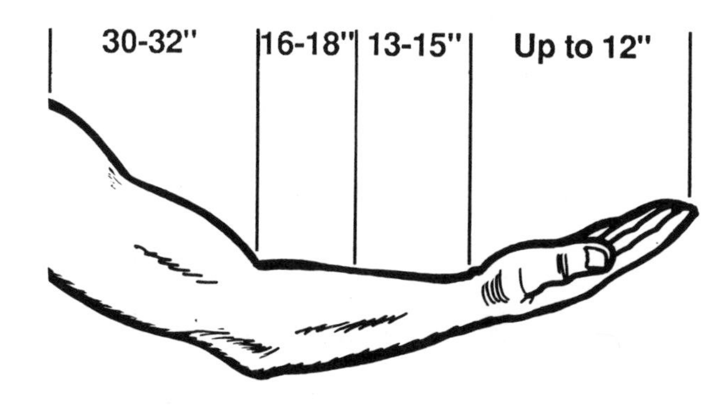

Chemical protective gloves are usually manufactured by the dipmolding or dipping process. In this process, forms or molds made of porcelain or metal in the approximate size and shape of a hand (Fig. 6-7) are immersed in a liquid "rubber" (Fig. 6-8). When the form is removed from the liquid, a coating of rubber is deposited (Fig. 6-9), which is further processed into a finished glove. Both unsupported and supported gloves are made this way except that for a supported glove, a fabric liner (Fig. 6-10) is placed over the form prior to dipping (Figs. 6-11 and 6-12).

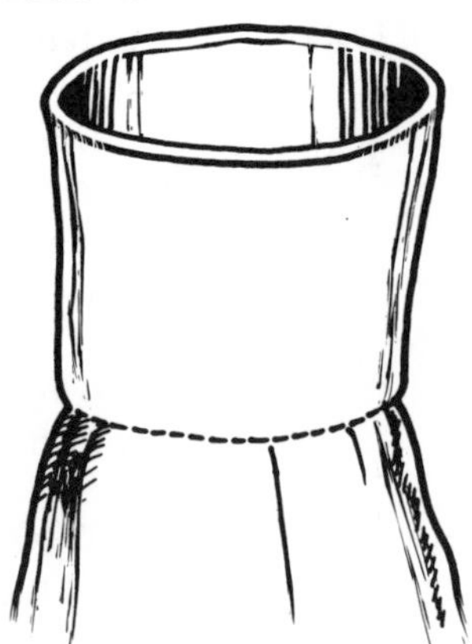

Gauntlet cuff: A coated straight cuff on a support glove

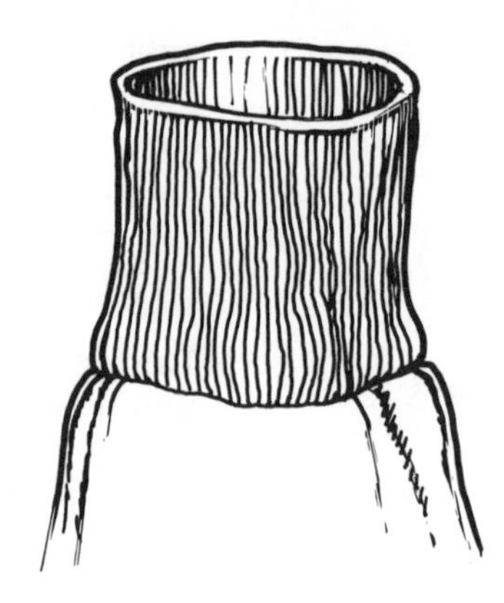

Knit-wrist cuff: A close-fitting cuff made from a soft, highly elastic knit fabric which is attached to the glove liner

Safety cuff: A cuff material which is attached to a supported glove with one side unsewn

Figure 6-6. Several common styles of supported glove cuffs.

courtesy Pioneer

Figure 6-7. Glove handforms on molds made from porcelain.

courtesy Pioneer

Figure 6-8. Glove handforms being dipped into a liquid rubber solution.

Figure 6-9. The coating of rubber deposited on the handform after dipping.

The liquid "rubber" solution can be either a latex system, a solvent-based solution, or in some cases, a plastisol. The latex technique is the most common process for the manufacture of chemical protective gloves. In a latex system, rubber particles are colloidally suspended in an aqueous dispersion. The latex system is compounded by conventional wet-mixing techniques (Fig. 6-13) with additives to enhance the desired physical properties. The non-water-soluble ingredients are added as dispersions (Fig. 6-14) or emulsions. Glove films are coagulated, normally using chemical means, and the thickness of the film is controlled by the specific gravity of the coagulant, the dwell time in the latex, and the latex solids content and/or viscosity. In most cases, lightweight and medium-weight unsupported gloves can be made in a single dip. This process is safer to use in the manufacture of gloves and less expensive than the solvent process.

In the solvent process dry rubber is mixed or milled with compounding chemicals (Fig. 6-15) into a homogeneous mixture (Fig. 6-16) and dissolved in the appropriate solvents. The correct dipping ratio of rubber solids is maintained by total-solids tests and viscosity checks. The forms are immersed in this rubber solution, then removed, allowing the solvent to evaporate. This leaves a thin film of dried rubber. Repeated dips are used to build up the desired thickness before the gloves are cured. This type of multiple dipping results in longer processing times, and uses volatile and flammable solvents instead of water, thus making this manufacture more costly and more hazardous than the latex process. Multiple-dip gloves, however, have certain superior chemical-resistance properties over latex gloves made from the same polymer.

In a plastisol system, a polymer such as polyvinyl chloride is dispersed in a liquid plasticizer (e.g., dioctyl phthalate, DOP). Forms are dipped, withdrawn, and allowed to drain. Glove thickness is controlled by the plastisol viscosity, form temperature, and drain time. The gloves are then "fused" at a high temperature and removed from the forms.

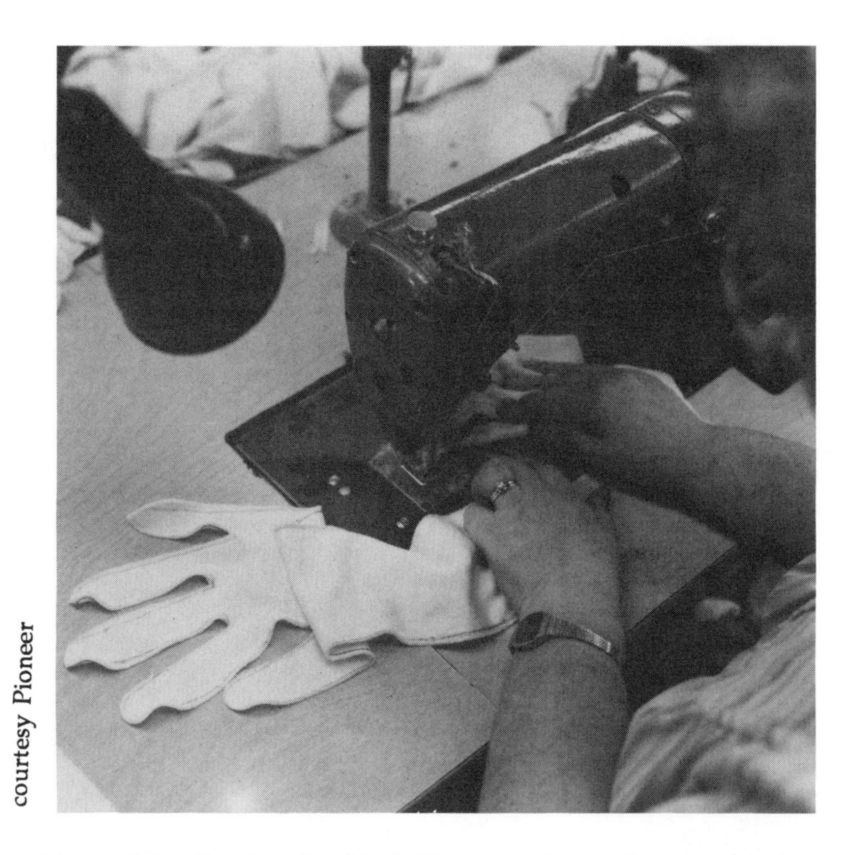

courtesy Pioneer

Figure 6-10. Sewing the fabric liners used to make supported gloves.

courtesy Pioneer

Figure 6-11. Mounting the support fabric liners on the hand-form before it is dipped.

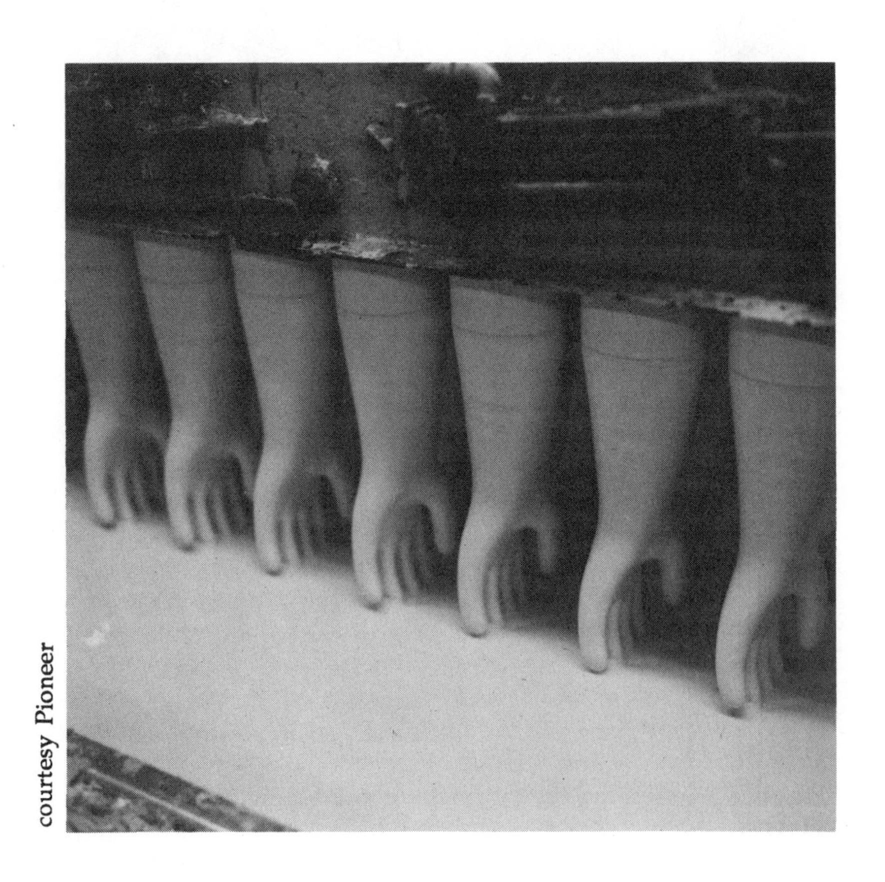

courtesy Pioneer

Figure 6-12. Dipping the handform into the liquid rubber.

Figure 6-13. Compounding the latex system using conventional wet-mixing tanks.

courtesy Pioneer

Figure 6-14. Facilities for grinding materials to produce latex dispersions.

courtesy Pioneer

Figure 6-15. Rolling mill used for dry mixing of rubber materials.

courtesy Pioneer

Figure 6-16. Milled dry rubber master batches ready to be dissolved in an appropriate solvent.

Foot Protection

Until recently, commercially available chemical protective boots were all made of PVC or rubber. With CPC users demanding more chemically resistant boots, manufacturers are developing a limited number of polymer blends that will be more resistant to certain workplace chemicals. Many problems are involved in using new polymer blends, though, due to the complicated injection molding process used to fabricate such boots. Care must still be exercised when boots come in contact with chemicals, because these boots can act as "chemical sponges" and expose the unsuspecting worker to hazardous chemicals.

The simplest boots are made using a single-stage injection molding process (Fig. 6-17). These products look like soft-sided rubber rain boots and are available in neoprene and butyl formulations (Fig. 6-18). Because of the one-stage process, the sole of the boot is made from the same polymer formulation as the sides, but is thicker. This means the wear and traction characteristics of the sole will not be optimum.

To provide a more durable and functional product, a two-stage injection molding process has been developed (Fig. 6-19). This allows a boot manufacturer the ability to produce a lightweight upper boot with an optimized long-wearing sole with good traction. The two-stage process also results in a better-fitting boot with improved chemical resistance (Fig. 6-20). These boots are available in PVC and PVC/nitrile formulations.

Hand-made boots are available in various foot sizes to provide better fit and comfort. These boots are made in stages with a large number of

Figure 6-17. Tingley's single-stage injection molding machine used for the manufacture of rubber overshoes.

courtesy Tingley

Figure 6-18. Examples of neoprene and butyl boots made from single-stage injection molding.

component parts (Fig. 6-21). The multicomponent structure of such boots is also prone to the "chemical sponge" hazard. Other styles of boots are available, made from neoprene and various rubber formulations.

Less-durable chemical-resistant footwear is available in the form of booties. Booties are made from inexpensive primary construction materials in the form of loose-fitting socks or shoe covers. When using this form of foot protection, one must realize that the construction materials have not been designed for significant wear properties. However, booties allow quick, easy, and inexpensive disposal of outer shoe coverings.

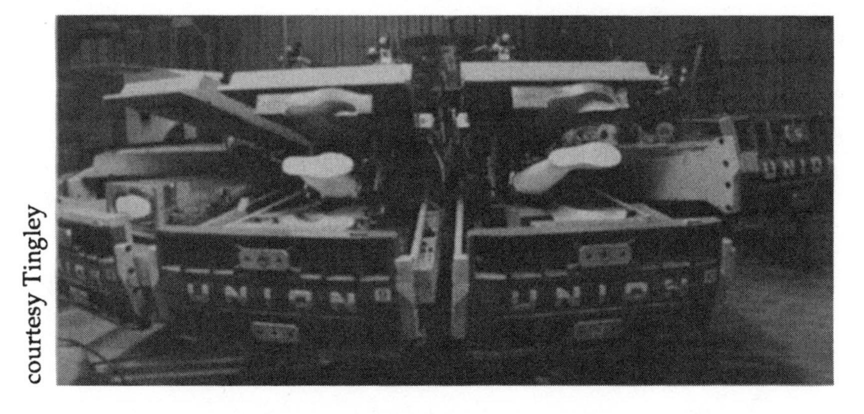

courtesy Tingley

Figure 6-19. Tingley's two-stage injection molding machine used for the manufacture of chemical-resistant boots.

Figure 6-20. Examples of chemical-resistant boots made from two-stage injection molding.

courtesy Tingley

Figure 6-21. Components of a hand-made rubber boot.

Body Protection

Body protection from the simplest form (the apron) to the most complicated form (the totally-encapsulating chemical protective suit) uses some or all of the other types of CPC.

The major component of body-protection equipment is the primary construction material. This material is a flexible plastic or rubber film, sheet, coated fabric, or laminate. A fabric can also be incorporated into the film to create a supported material, which is stronger and more resistant to tear and puncture. Nylon, dacron (polyester), Nomex, and fiberglass are used as supports in fabrics. Figure 6-22 illustrates the typical construction cross section of a supported fabric with a butt seam. Table 6-5 lists the commercially available supported fabrics for CPC manufacture.

To better illustrate a coated-fabric process, we use the example of how PVC-coated fabric is made. Figure 6-23 illustrates a typical PVC-transfer coating line. The PVC-transfer coating procedure begins when a silicone-treated release paper is attached to the first of a series of rolls, called an accumulator. The accumulator provides the

Table 6-5. Primary construction materials of commercially available TECP suits.

Primary Construction Material[a,b]	Ensemble Vendor[c]
Butyl/Nylon/Butyl	Kappler Safety Group Fyrepel Products Trelleborg
Butyl/Polyester/Chloroprene	Mine Safety Appliances (MSA)
Chlorinate Polyethylene	ILC Dover Standard Safety Equipment
Chlorobutyl/Nomex/Chlorobutyl	Arrowhead Products ILC Dover
Chloroprene/Nylon	National Draeger
Polyvinyl Chloride/Nylon	Kappler Safety Group
Polyvinyl Chloride/Polyester	Standard Safety Equipment
Polyvinyl Chloride/Nylon/ Polyvinyl Chloride	Fyrepel Products National Draeger Trelleborg Wheeler Protective Apparel
Teflon/Fiberglass/Teflon	Chemical Fabrics Corporation
Viton/Nylon/Chloroprene	Mine Safety Appliances National Draeger
Viton/Butyl/Nylon/Butyl	Trelleborg
Viton/Polyester/Viton	Fyrepel Products

[a] The primary construction materials are described with the external surface first and the surface facing the wearer last.

[b] Available in thickness ranges from 8 to 28 mils (one mil is equivalent to 0.001 in. or 0.025 mm).

[c] See the appendix volume for the address and telephone number of each vendor.

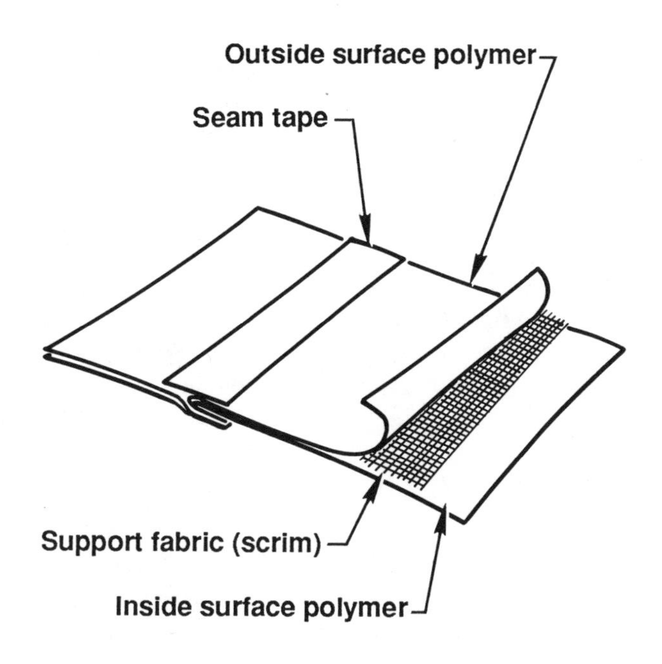

Figure 6-22. Typical construction cross section of a supported fabric with a butt seam.

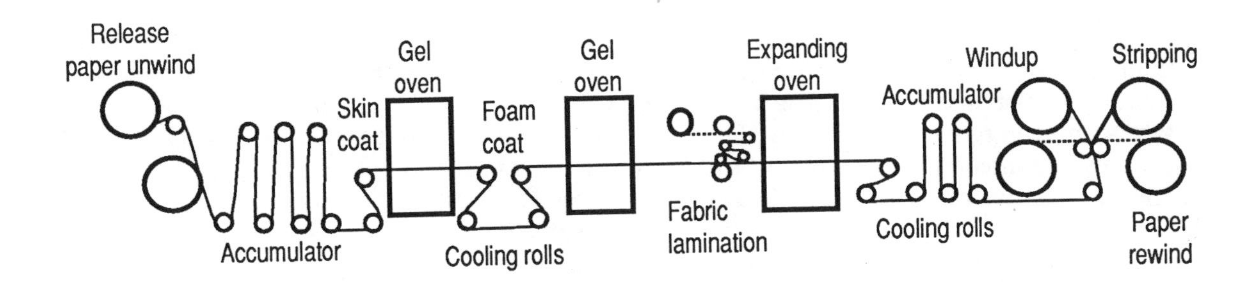

Figure 6-23. Schematic of a typical PVC transfer coating line.

delay time needed to splice a new roll of paper, or a different type of paper, without shutting down the line. After the paper leaves the accumulator, the first coat—or skin coat—is applied to the release surface of the paper, using a common coating technique such as knife-over-roll, knife-over-blanket, reverse roll, or rotogravure. This step is followed by processing in a drying or gelling oven. The paper is then routed over cooling rolls to the second coating station, where the tie coat is applied to the skin coat. The fabric is then brought into contact with a wet tie coat by a pair of laminating rolls. The fabric-resin construction is dried in a three-stage oven and cooled and stripped from the release paper. The stripping process involves two rolls, one of which is used to store the finished product, the other to store the release paper for reuse.

Careful surface examination of the primary construction material is important to determine if surface imperfections exist, as well as pinholes. Such imperfections can affect the performance of the CPC material. If one looks at supported butyl rubber fabric under a 10 kV × 20 power electron microscope, one begins to see surface imperfections (Fig. 6-24). At greater magnification, 15 kV × 20 power, one can clearly see the imperfections, the support fabric, and the thickness of the butyl coating (Fig. 6-24). At an even higher magnification, the microstructure of the butyl coating can be seen (Fig. 6-25).

Fabric can be joined in two ways to form either a lap seam or a butt seam(Fig. 6-26[a] and [b]). The choice of seam depends on the material type, required seam strength, and desired physical appearance of the final product. The joining of the two seams can be done by welding (Fig. 6-26[a] and [b]) or stitching (Fig. 6-26[c]). Stitching is normally done with nylon thread in a double seam to provide a stronger seam. Since the stitching process creates holes in the primary construction material, these holes must be sealed to provide either a liquid-proof or a gas-tight garment. Tape or strapping is cemented or welded over the sewn seam (Fig. 6-26[d]). CPC seams should be carefully evaluated to make sure they are strong enough and sealed properly.

Various chemical-resistant coats and pants are available to prevent direct skin contact with low to moderately toxic materials. These coats and pants are typically made from supported PVC, neoprene, or SB rubber primary construction materials. Many two-piece outfits are available as industrial rain gear. Also available are limited-use/disposable models fabricated from polyethylene-coated or Saranex-laminated Tyvek as the primary construction material. In better models, care is taken to provide high-quality liquid-proof seams. Suit closures are snaps and zippers with a storm fly to prevent most liquid penetration.

Splash suits consist of a chemical-resistant coat and pants, or coveralls equipped with a hood. Additional primary construction materials available in splash suits are chloroprene, polyurethane, butyl rubber, and a blend of nitrile rubber and polyvinyl chloride. Splash suits are designed to prevent direct skin contact with liquids, but provide little or no vapor protection. This lack of complete liquid or gas protection is due to closures and interfaces that are not gas- or completely liquid-tight. The use of splash suits must therefore be evaluated carefully.

Limited-use/disposable CPC presents an interesting example of the manufacturing process necessary to construct a competitive product. Most CPC manufacturing processes involve labor-intensive operations that must be managed carefully to control costs. To illustrate this type of control, Durafab, Inc. has supplied representative pictures of their manufacturing facilities. In Fig. 6-27, the large cutting table is shown where 150 feet of various Tyvek primary construction materials are rolled out and stacked 400 layers high. Patterns are laid out manually on the stacked material and cut, using an industrial saber saw (Fig. 6-28). The patterns are laid out with great care to minimize waste.

Some other manufacturers are using computer-aided pattern-tracing techniques to minimize the waste. All waste is collected and recycled as scrap plastic. The various clothing components are transferred to the sewing room where they are assembled with standard and modified sewing techniques (Fig. 6-29). Specially constructed seams are ultrasonically sealed to improve their strength and liquid tightness (Fig. 6-30). Alternate seam-closing techniques have been incorporated in Kappler products to make their seams liquid proof (Fig. 6-31).

Limited-use/disposable CPC provides an important type of protection and can be used to minimize employee exposure to hazardous chemicals at a reasonable cost. High levels of protection should not be expected from these types of garments, though, since they are not designed for such an application.

Low magnification

Moderate magnification

Figure 6-24. The surface of supported butyl rubber at low and moderate magnifications.

Figure 6-25. Supported butyl rubber surface at high magnification.

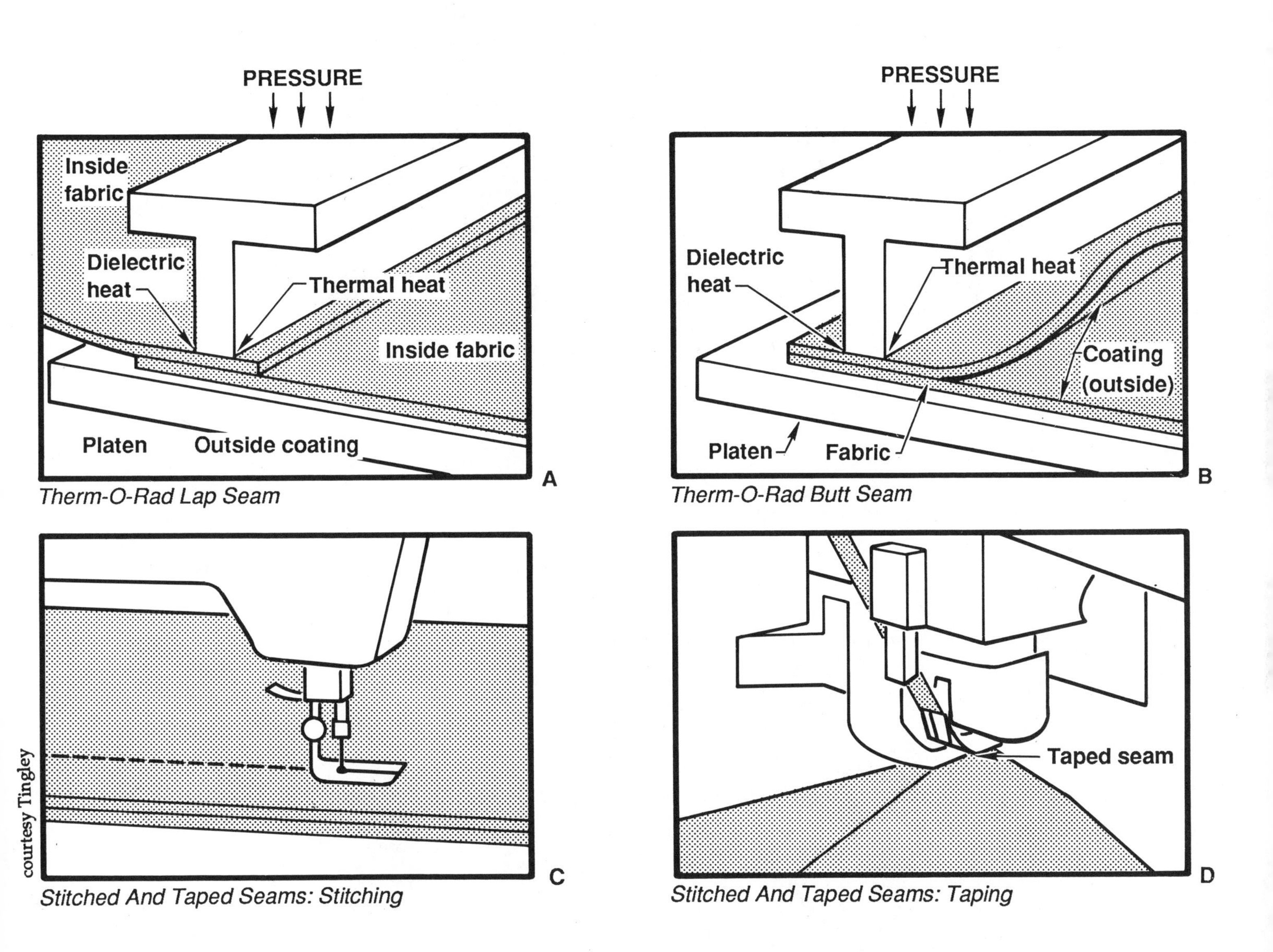

Figure 6-26. Various seam types and sealing techniques used to fabricate CPC.

courtesy Durafab, Inc., Cleburne, TX

Figure 6-27. Durafab's 150-foot cutting table with automatic fabric spreader.

courtesy Durafab, Inc., Cleburne, TX

Figure 6-28. Durafab's process for cutting CPC components from their primary construction material (approximately 400 layers).

courtesy Durafab, Inc., Cleburne, TX

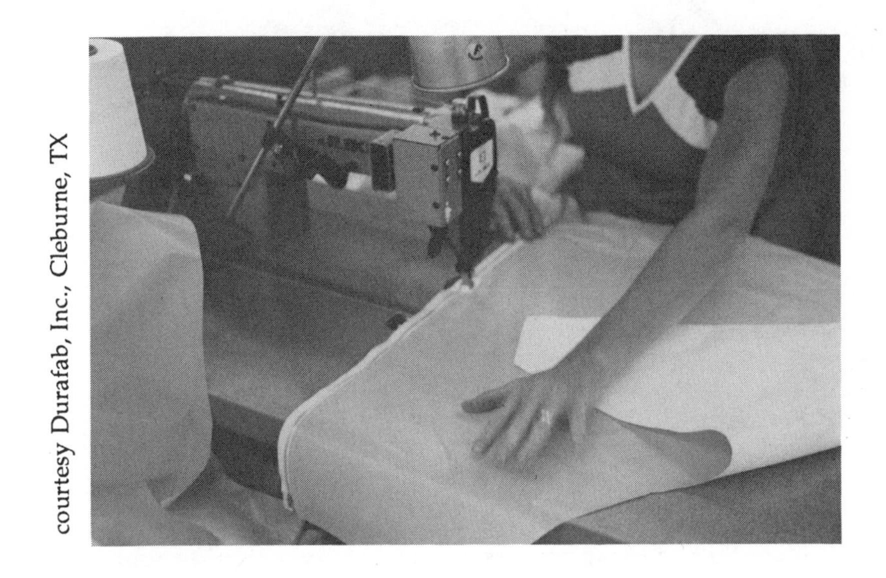

Figure 6-29. Durafab's garment-assembly process, using conventional sewing techniques.

courtesy Durafab, Inc., Cleburne, TX

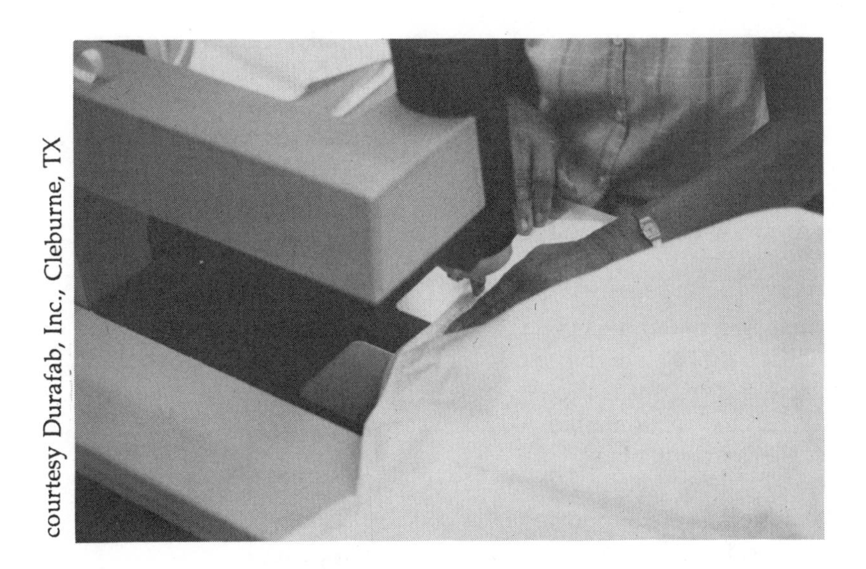

Figure 6-30. Durafab's garment-assembly process, using ultrasonic sealing techniques.

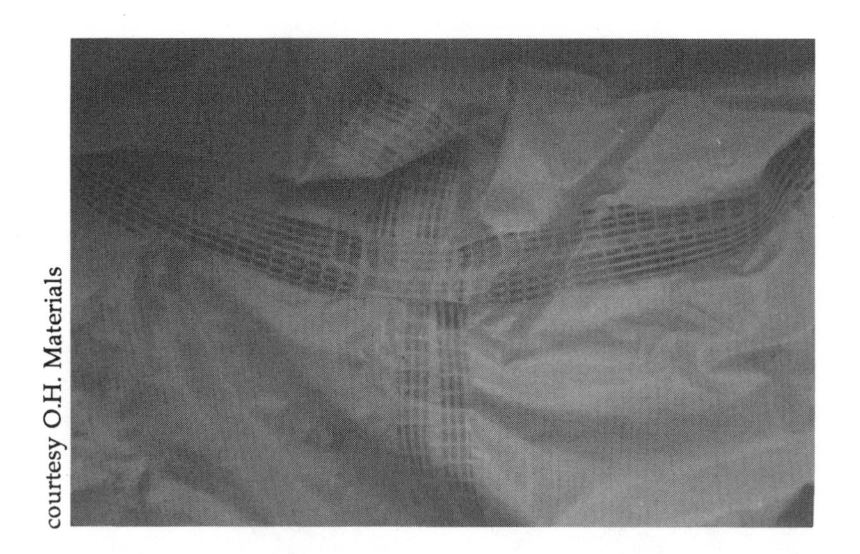

courtesy O.H. Materials

Figure 6-31. Kappler's seaming process, used to produce liquid-proof seams.

Totally-Encapsulating Chemical Protective Suits

The totally-encapsulating chemical protective (TECP) suit combines all of the various CPC types into one full-body covering. Specialized components such as vent valves and gas-tight closures are added to the suit for proper operation. Table 6-6 identifies the components of a TECP suit that prevent chemical exposure to the wearer.

The primary construction material is the major component of a TECP suit. The high performance of this material and its seams is therefore essential for a TECP suit to function effectively in adverse chemical environments.

Since many of the materials are supported composites, each is described from left to right, identifying the various layers from the outside surface to the inside surface. For example, "butyl/polyester/chloroprene" indicates that butyl rubber is the external surface and the chloroprene is the internal surface of the suit. Between these two polymer films is a polyester support fabric.

Table 6-6. The components of a TECP suit that prevent chemical exposure to the wearer.

- **Primary construction material**
- **Seams**
- **Closures**
- **Gloves**
- **Visors**
- **Boots**
- **Vent valves**

To aid the user in identifying TECP suit wear, colored indicator layers are sometimes incorporated into the primary construction material. This allows the repair staff and the user to identify suit areas that require repair. In some cases, the TECP suit may need to be discarded.

The closures used on TECP suits are key components in keeping the suit gas-tight. Three types of gas-tight TECP suit closures are currently used—pressure-sealing zippers, Ziploc closures, and a thermally sealed coffer dam flap (Fig. 6-32).

The pressure-sealing zipper is designed such that the teeth of the zipper protrude only slightly through a plastic/rubber-coated fabric. When the zipper is closed, the teeth lock, tightly pressing the plastic/rubber on the two sides of the zipper and forming a gas-tight seal. Current pressure-sealing zippers are manufactured using either chloroprene, PVC-coated nylon, or PVC-coated polyester fabric. Chloroprene is more commonly used because of the poor aging characteristics of PVC. The teeth of the zipper are normally nickel silver equipped with a brass slider, but stainless-steel zippers can be specially ordered. These zippers are expensive and require careful use to assure their continued operation.

The Ziploc sealing channel is an effective and inexpensive type of gas-tight closure. Extended interlocking chlorinated polyethylene strips are incorporated into the TECP suit. A conventional zipper is used to close the suit entry and provide the strength to keep the entry area closed. The Ziploc-style channels are closed on the outside of the conventional zipper, making a gas-tight seal.

The third type of sealing mechanism relies on a thermally welded seam made on coffer dam flaps around the suit entry area, after the wearer has donned the TECP suit. This type of seal is nonreusable because it is broken to permit the wearer to exit the suit. By designing large coffer dam flaps, however, several seals can be made on the same TECP suit. Other variations in TECP suit-closure design are the provision of a cover flap to minimize direct chemical contact to the actual closure and the length of the entry area.

The actual location of the entry area determines whether the wearer can don the TECP suit without assistance. If the closure is located on the front of the suit, it is possible for the wearer to seal it without help. If the closure is located on the back of the suit, a second person is required to operate the closure. If a thermally welded closure is used, a second person is required, along with a source of 110 V ac power.

Since the TECP suit is only as strong and as chemically-resistant as the weakest seam, care must be exercised in the joining of the closure device itself to the primary construction material. Careful testing is required to assure this has been done properly.

Gloves and boots are attached to the extremities of the TECP suit to protect the wearer's feet and hands from chemical exposure. For the suit design to be effective, the boots should fit the wearer as well as the gloves. Several different designs are available to meet these performance requirements.

Originally, several TECP suit manufacturers incorporated gloves as a permanent part of the suit. This design was found to be unacceptable due to fit, repair time, and decontamination requirements. Now, most suit manufacturers provide removable gloves as part of their design. The gloves are attached to the suit by concentric rings, ring/clamp assemblies, or connect rings (Fig. 6-33).

Concentric Rings—This interface is based on two tapered rings about two inches wide and four inches in diameter. One ring is slightly smaller in diameter than the other and fits snugly into the larger one. The larger ring is positioned inside the end of the garment sleeve. The smaller ring is placed inside the glove at the cuff. The glove is then pushed down the inside garment sleeve from the shoulder. As the glove passes through the sleeve opening, the smaller insert is pressed inside the larger one, forming a compression seal.

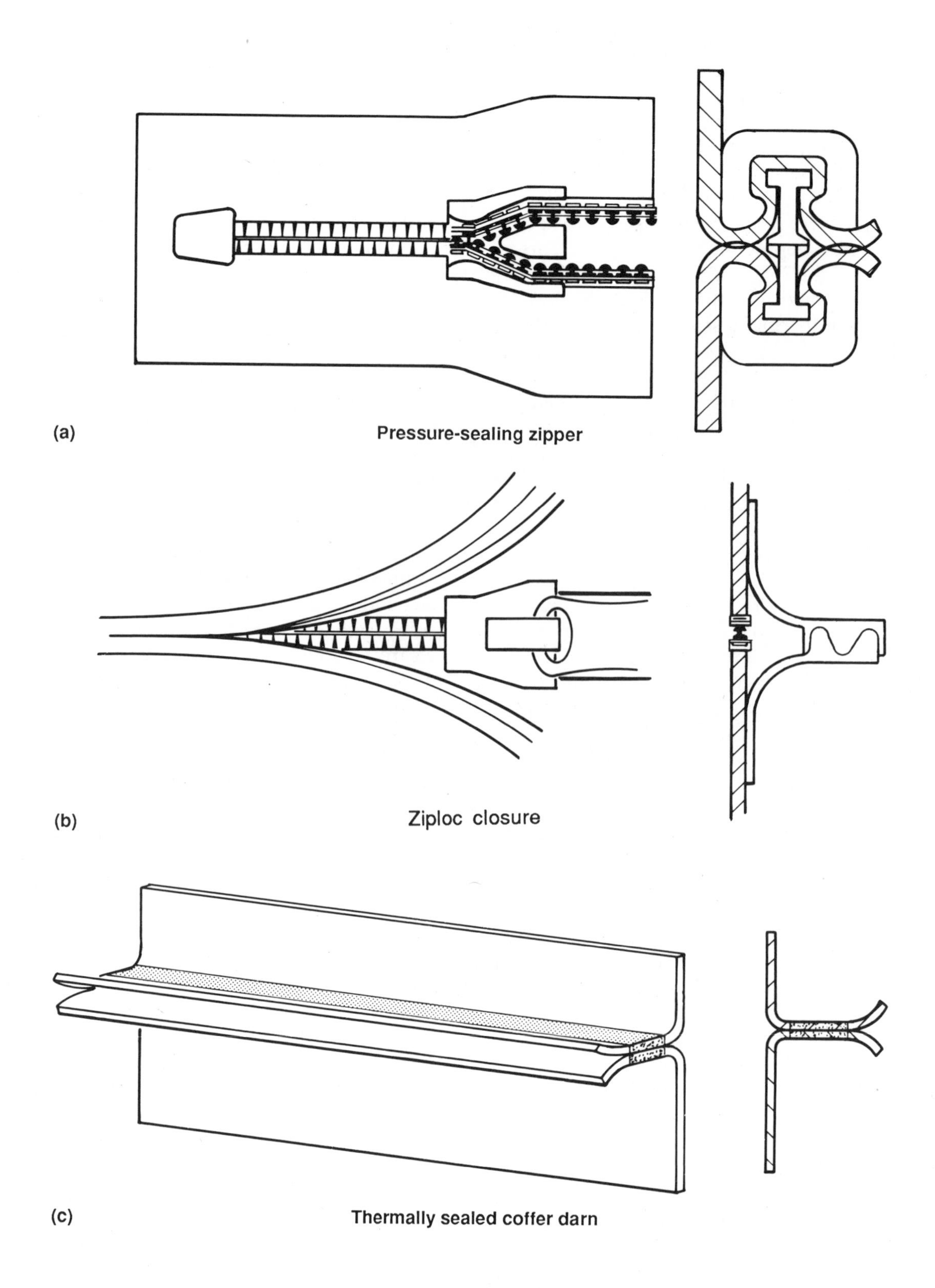

Figure 6-32. Currently used TECP suit gas-tight closures.

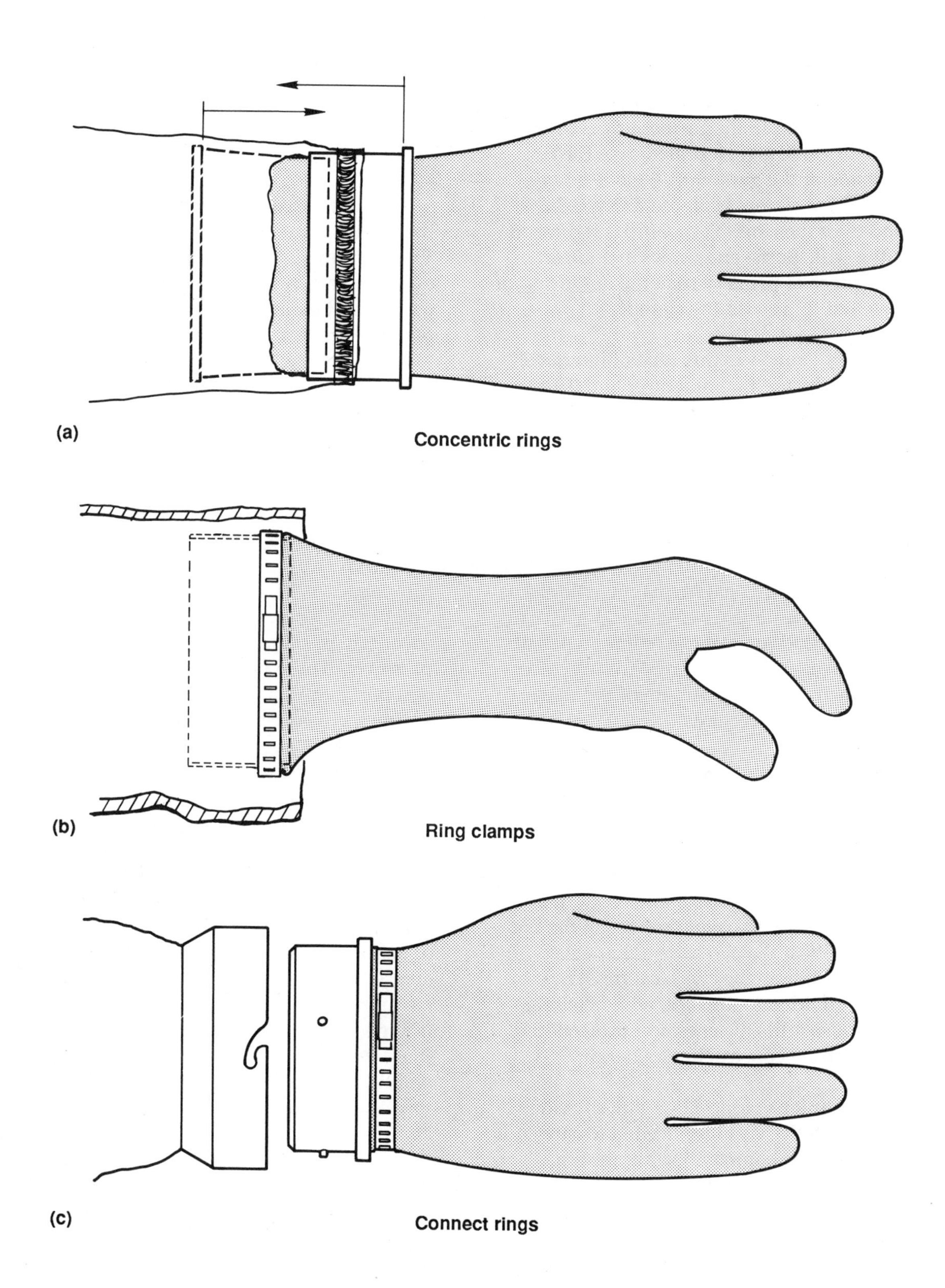

Figure 6-33. Glove attachment techniques for TECP suits.

Ring/Clamp—This interface uses a support ring in the sleeve and a clamp around the glove. The ring, about four inches in diameter and one inch wide, is placed in the garment sleeve near the cuff. The glove is then pulled over the ring on the outside of the garment. A glove clamp, similar to a hose clamp, is placed around the glove and over the ring. The clamp is tightened to form the seal. Some ensembles include an additional piece of primary material at the end of the sleeve that folds over the clamp.

Connect Rings—In this type of interface, rings installed by the manufacturer at the glove and the sleeve cuffs are mechanically joined using an O-ring seal to create a gas-tight closure.

Selecting the correct glove to be attached to the TECP suit depends on the field application. A careful evaluation of cut and puncture strength, chemical permeation, manual dexterity, temperature, and other glove qualities should be made. A second pair of gloves (overgloves) can sometimes be used to meet the requirements of the job effectively. Some suit designs also incorporate a splash cuff at the bottom of the sleeve to pull down over the overglove.

Chemical-resistant boots can be permanently attached to the TECP suit or can be removable. Some suits may have a sock foot made from the primary construction material, allowing the wearer to use any commercially available boot. A splash cuff is normally incorporated in the sock-foot design to cover the tops of the boots from splash. With the permanently attached or removable boot, only the chemical resistance of the boot is provided to the foot, while the sock-foot design also adds the chemical protection of the suit's primary construction material.

The visor lens of the TECP suit provides the wearer with the visibility needed to carry out various tasks. The composition and testing required to select the correct visor material was discussed above (see *Eye and Face Protection*). Fogging from moisture condensation inside the suit can create a significant visibility problem. Antifogging treatments are available for coating the lens interiors, but they are only partially effective. New lens developments might eliminate this problem and may provide increased chemical resistance along with improved visibility.

Vent valves are used in a TECP suit to let the exhaled or cooling air escape from the suit. This minimizes ballooning without allowing exterior air to reenter. Two types of valves are used: the standard vent valve (Fig. 6-34) and the positive-pressure vent valve (Fig. 6-35).

The standard vent valve consists of a polymer flapper diaphragm attached to a valve seat. The valve seat is connected to the TECP suit, and a splash pocket covers the outside of the valve. The rubber diaphragm opens at a very low pressure (cracking pressure ~ 0.2 in. water, gauge) and vents the suit. This valve design was used widely in early suits and had few reported problems. As TECP suit design became more sophisticated, several manufacturers began to incorporate positive-pressure vent valves into their products. These valves operate in a similar manner to the standard vent valve, except that a spring increases the cracking pressure to as high as 3 in. water, gauge. This increased pressure causes the suit to balloon, but also provides a positive ballast pressure. The ballast pressure prevents the suit interior from becoming negative during various body motions, thus minimizing the chance for reentrainment. A positive pressure of 1–1.5 in. water, gauge, appears to be a good compromise between the negative effects of ballooning and the positive effects of increased interior pressure. Experiments are currently being conducted to determine if there is a measurable performance difference between these two vent valve designs. Until definitive results become available, the purchaser should rely on the manufacturer's recommendations and previous experience.

Several additional components allow the TECP suit wearer to function properly: a breathing-air supply system, a communication system, and a cooling system. The breathing-air system is one of the most important parts of a

Figure 6-34. The standard vent valve used in TECP suits.

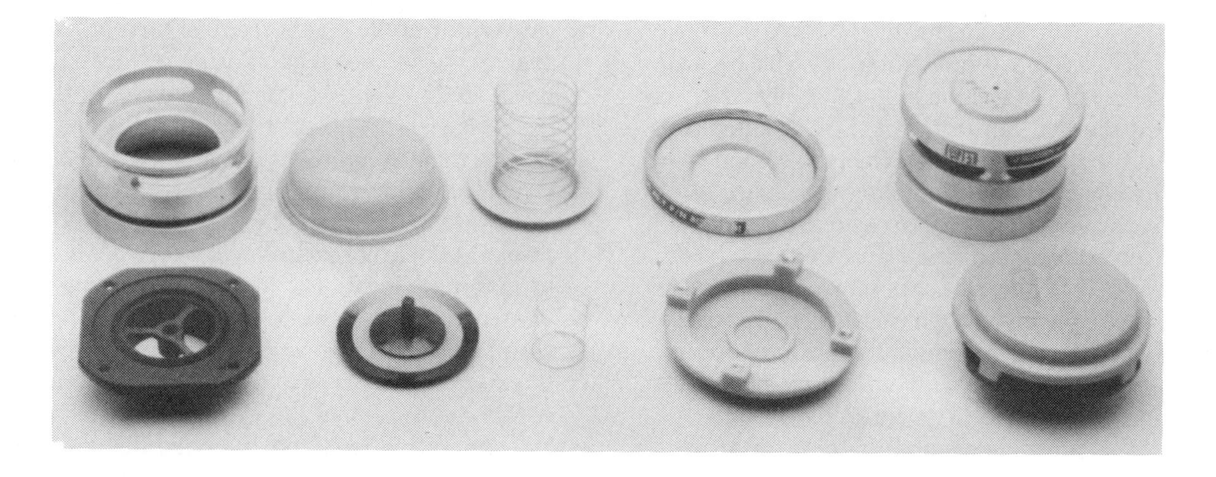

Figure 6-35. The positive-pressure vent valve used on TECP suits.

TECP suit. Two styles of TECP suits are currently available, each containing the same components, but they differ in the location of the positive-pressure SCBA (Fig. 6-36). Type I suits completely enclose the SCBA and the wearer, while Type II suits completely enclose the wearer but locate the SCBA on the outside of the suit where it could be exposed to a hazardous chemical environment. To date, no studies have shown any difference in performance between the two suit styles. Both the Type I and Type II styles have uses in hazardous material responses. The low profile offered by the Type II design may enable certain jobs to be done more effectively and safely. On the other hand, a real concern about decontamination would certainly exist if the Type II TECP suit and SCBA became contaminated.

A positive-pressure airline respirator with an escape bottle can also be used with TECP suits. This form of respiratory protection provides a longer duration of work time with less weight from the breathing apparatus. Normal breathing air is provided by either a breathing-air compressor or a compressed-air-bottle bank. Air is supplied to the worker through a compressed-air umbilical hose that must be used carefully to prevent contamination of the breathing air. The escape bottle provides an independent source of breathing air to allow the wearer to escape hazardous situations. The duration of use for these bottles is normally five minutes.

The NASA John F. Kennedy Space Center has developed a unique respiratory system—the Environmental Control Unit (ECU), which uses liquid breathing air. The ECU is designed to be used exclusively with the Kennedy Space Center's propellent handler's ensemble to provide both breathing air and worker cooling.[6-8] The ECU is designed to operate for 120 minutes and to supply air at a temperature of approximately 75°F with a relative humidity of about 50%. This is the only currently used TECP suit (the propellant handler's ensemble plus the ECU) that addresses both breathing air and worker cooling in an effective manner. Unfortunately, the use of liquid air as breathing air is not a common practice, and the ECU and propellant handler's ensemble are not commercially available. The NASA Kennedy Space Center is, however, evaluating a modified one-hour ECU to be submitted to the National Institute for Occupational Safety and Health (NIOSH) for certification.

Selection of the correct type of respirator for specific uses of TECP suits requires a careful analysis. Normally, weight and duration are the two factors that determine the type of respirator selected. Due to the variation in size and fixture requirements, the compatibility of respirators with specific TECP suit models should be confirmed with the manufacturer.

Communication between wearers of TECP suits is important in a work environment. Because of the combined limitations of the SCBA and the TECP suit, some type of electronic amplification is necessary. Three types of electronic amplification systems are presently available: voice amplification, hard-wire systems, and wireless devices. Voice amplification systems use a microphone attached by wire to an amplifier speaker box mounted somewhere on the wearer, normally on the belt of the suit. This system is most effective when work is being carried out in a closed area, and no remote communication is required. Hard-wire systems use microphones and headsets physically connected by wire between the workers or back to a switchboard. The switchboard allows the workers to communicate with remote locations or to each other. This system works best when airline respirators are used, so that the wire can be attached to the breathing-air umbilical hose. Wireless systems are the most convenient but also the most expensive. They are essentially a portable radio transmitter/receiver that allows the worker to transmit voice communication between workers or to remote locations. They also offer the convenience of easy remote monitoring to allow for simple job-progress analysis and quick response to emergencies. Due to the complicated nature of these communication systems, care should be exercised when choosing one. Regular maintenance and repair should also be planned, along with the purchase of several backup units.

Figure 6-36. Type I and Type II TECP suits.

Because of the lack of evaporative cooling in most TECP suits, along with an increase in workload and exertion, heat stress from wearing a TECP suit and related equipment is recognized as a significant limitation to the wearer. Various methods are available to reduce the heat stress encountered by the TECP suit wearer.

Passive systems such as vests containing ice, dry ice, frozen gels, or other heat sinks can be worn. These systems require equipment for pre-freezing the heat-sink contents before use. Typically, such systems have a service life of one to four hours, depending on the workload and the external temperatures. These systems are usually heavy and cumbersome, and the wearer has no control over the cooling rate or cooling distribution.

Powered systems in which a chilled fluid is pumped through tubes in contact with the head, neck, chest, and other body regions can be worn. The circulated fluid may be air, water, or some other heat-transfer medium. The heat sink in this method is ice, dry ice, or frozen gels. These systems are battery powered and typically have operating periods of one to four hours. In some systems, the temperature can be controlled by the wearer. This method also requires equipment for pre-freezing the heat-sink contents before use.

Umbilical air-cooling systems that distribute air to the head, neck, and other body regions through a tubular manifold system can be used. The cooling air is exhausted through the suit vent valves. Air-flow rates can be controlled by the wearer, and cooling is generated by sweat evaporation. Umbilical air-cooling systems in which pressurized air enters the ensemble through an expansion valve (i.e., vortex tube) are also available. The level of noise from the expansion restricts the proximity of the air inlets around the head, but the wearer can control the rate of cooling by adjusting the air flow.

Powered cooling systems based on swing, Stirling cycle, and conventional compressors are being examined for commercial use. To date these systems have not proven practical because of their weight. However, new developments in high-efficiency, high-storage-capacity batteries may change this situation.

In addition to the above systems, cooling can be achieved from the evaporation of water sprayed over the surface of the ensemble. Cooling systems are important, but care needs to be exercised to make sure the increased weight and reduction in maneuverability are acceptable.

TECP suits must be initially tested to assure that the specific suit design is functional and routinely tested to assure that the suit remains operational. Johnson and Stuhl have proposed four types of tests to fulfill these requirements.[9]

- A quantitative test using both a gas and an aerosol challenge agent is being developed to measure the leakage of test agents into the suit. Figure 6-37 shows the NASA TECP suit being evaluated in a man-test chamber. A suit protection factor (outside concentration of the test agents divided by the inside concentration) would be generated by this test.

- A worst-case chemical exposure test that measures the suit performance in high concentrations of hazardous materials has been proposed and demonstrated.[10] Figures 6-38 and 6-39 show the US Coast Guard's new Teflon-coated Nomex TECP suit being tested in a hydrogen fluoride cloud at the US Department of Energy's Nevada Test Site. Figure 6-40 illustrates a remote worst-case chemical exposure test concept using a mannequin to test TECP suit integrity.

- A test method is available to measure the leak rate of a pressurized suit. This test method is the American Society for Testing and Materials (ASTM) F1052-87, "Standard Practice for Pressure Testing Gas-Tight Totally Encapsulating Chemical Protective Suits."[11] Figure 6-41 illustrates how this test is carried out. Because of the test's simplicity, it can be used to check TECP suits in their design process and during regular use.

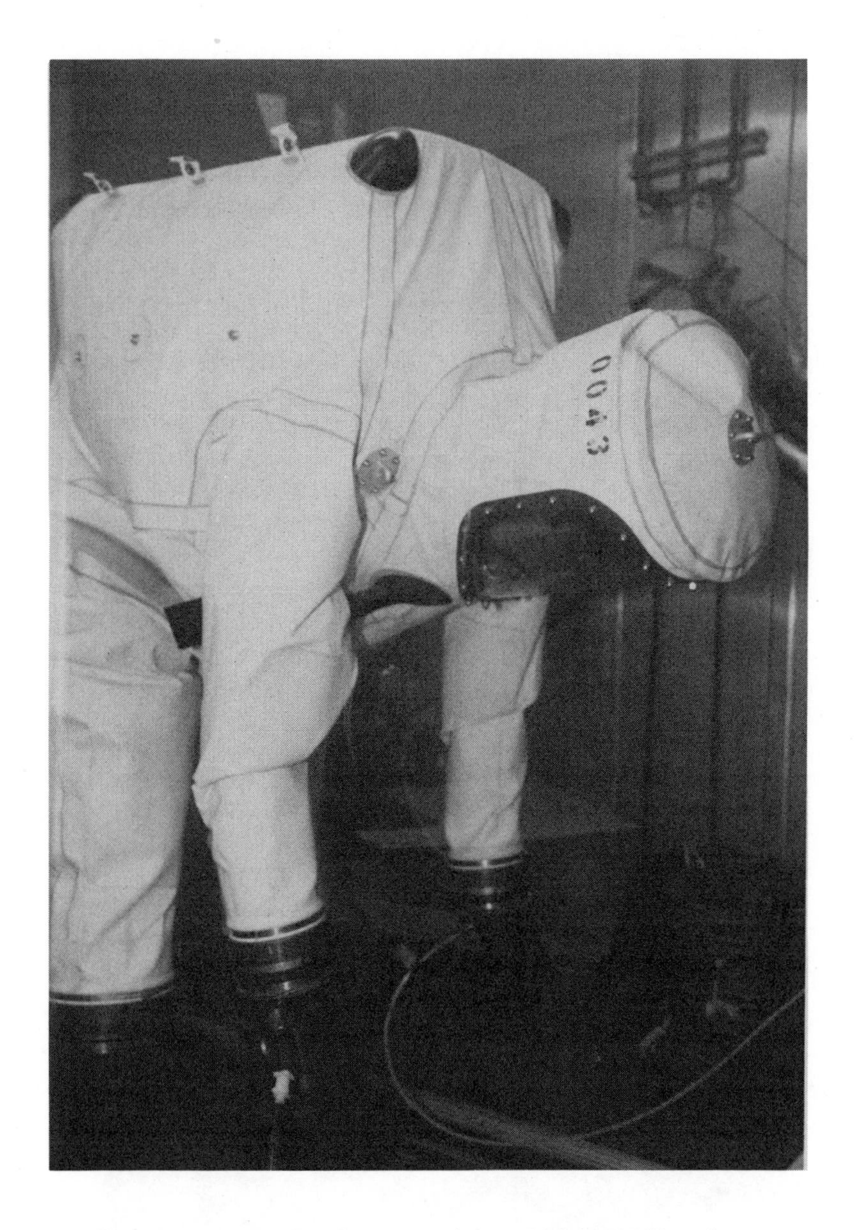

Figure 6-37. Evaluation test of the NASA TECP suit.

Figure 6-38. Hydrogen fluoride test of US Coast Guard TECP suit (Nevada Test Site).

Figure 6-39. US Coast Guard suit, prepared for HF tests.

• A proposed chemical leak-rate test is being developed that will use a known ammonia concentration generated in a test room and a worker wearing the TECP suit into the room and performing a series of prescribed light exercises. The interior of the suit is monitored for ammonia after the subject exits the test room. A 'protection factor' is calculated as described for the quantitative test, above.

An additional ASTM standard practice has been finalized by Committee F23 on Protective Clothing that addresses the measurement of qualitative properties of protective suit ensembles. It is ASTM F1154-88, "Standard Practice for Qualitatively Evaluating the Comfort, Fit, Function, and Integrity of Chemical Protective Suit Ensembles."[12] By employing these five TECP suit tests in the various stages of suit development and field use, a high degree of suit safety can be assured as well as reasonable fit, function, and comfort.

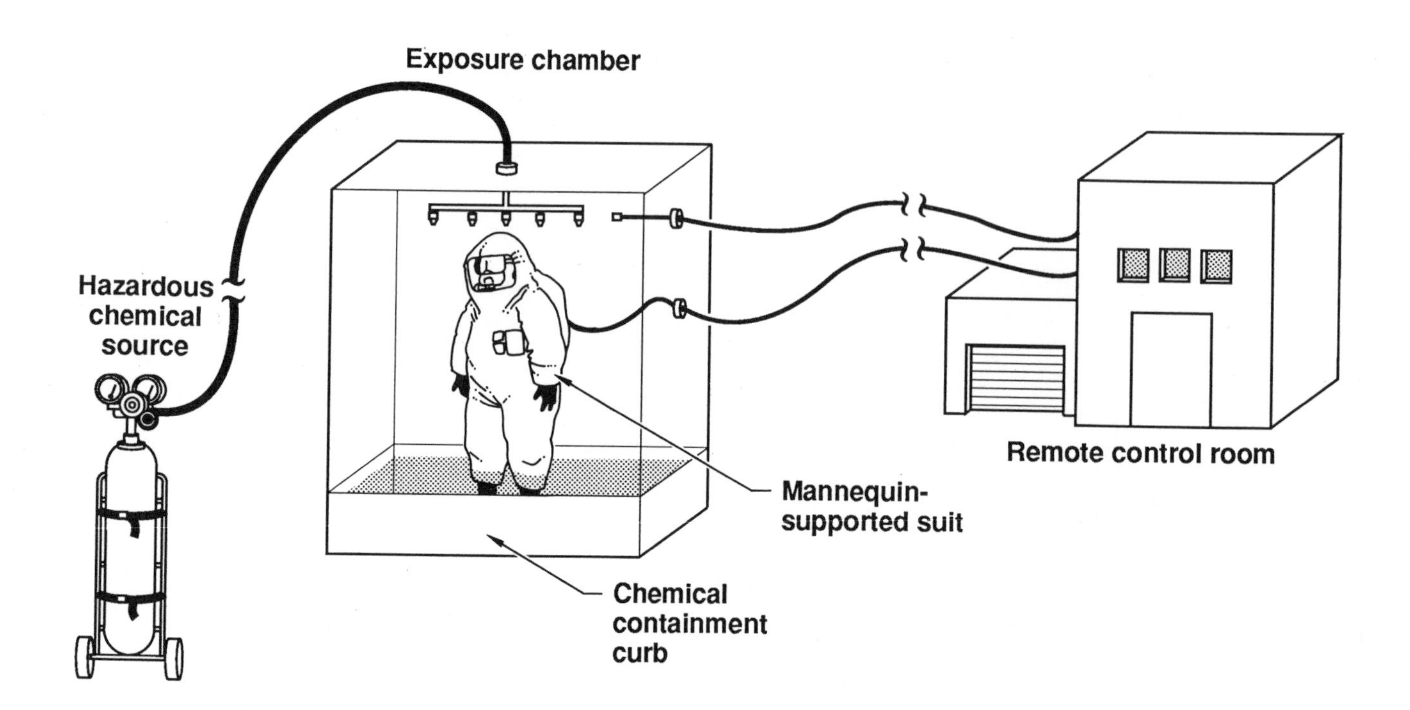

Figure 6-40. Remote worst-case chemical exposure test. The suit under test is supported by a mannequin.

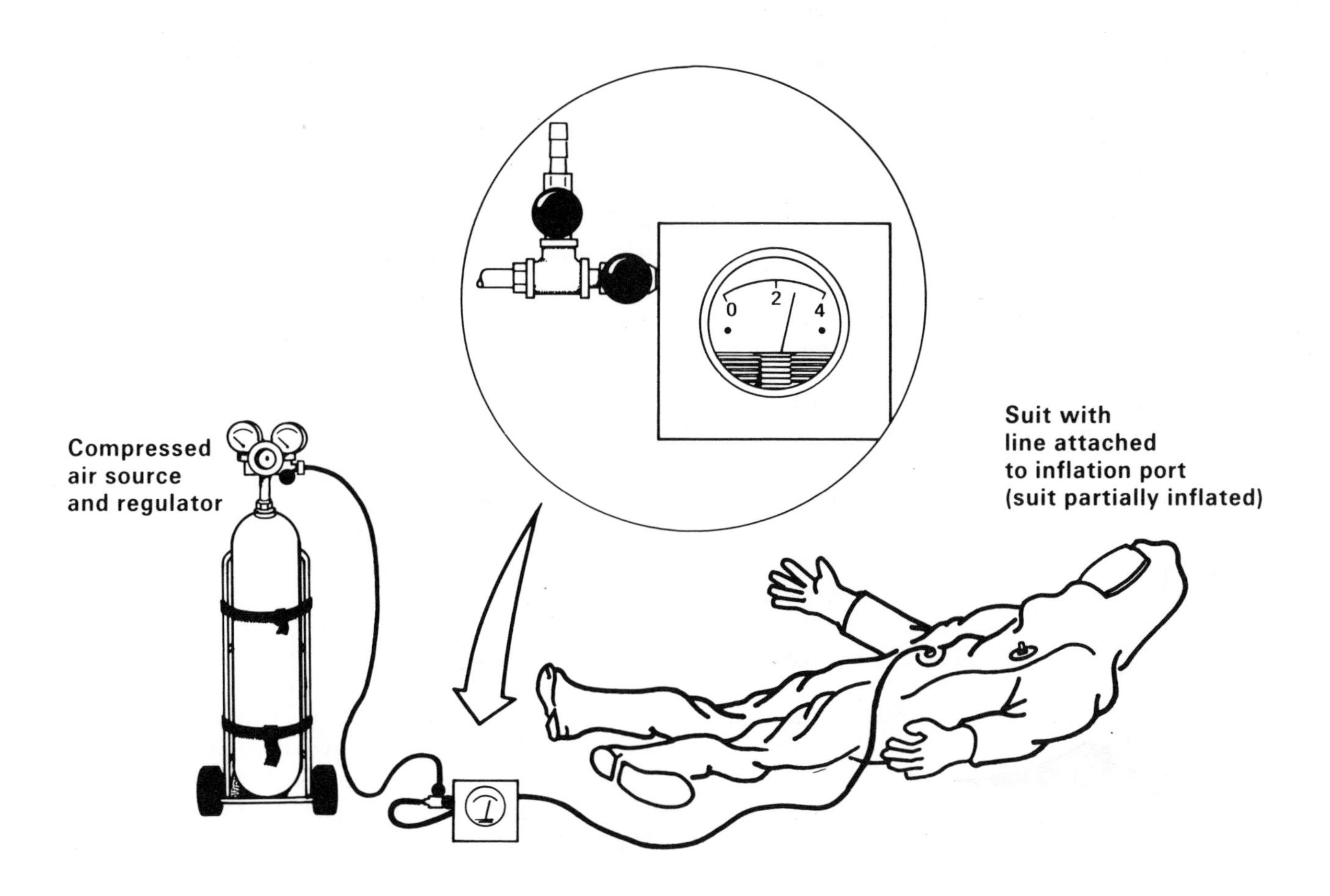

Figure 6-41. ASTM pressure testing of a gas-tight TECP suit.

Summary

A wide variety of chemical protective clothing is now available for the workplace. Manufacturers are paying special attention to improving chemical resistance, dealing with heat stress, offering multiple sizes, and addressing womens' needs, as well as style, comfort, and worker acceptability.

The major types of CPC are: eye and face protection, head protection, hand protection, foot protection, partial body protection, and complete body protection. The protection offered by each of these major types of CPC can be understood better if they are analyzed as a component of a TECP suit.

The need to know limits of eye and face protection was illustrated by a dramatic TECP suit lens failure at Benicia, California. This failure showed that selection of eye protection simply for visibility is not good enough. Chemical resistance must also be considered.

Head protection is not normally a stand-alone consideration when choosing CPC. Protective devices such as splash hoods are available, but they are normally used in conjunction with two-piece whole-body suits or are completely incorporated into a TECP suit.

Hand protection has received much attention in recent years because of the many different tasks involving the hands in direct contact with toxic and hazardous chemicals. Gloves are available in a variety of polymers and thicknesses that provide good resistance to most of these chemicals. When choosing the proper chemical-resistant glove to maximize its performance in the workplace, one must consider whether the glove is supported or unsupported, has an acceptable supported liner design, flocked or unflocked glove interiors, whether it was latex solvent or plastisol-dipped, and the length of the glove.

Improved polymer systems and molding techniques are being developed that can be used to produce better chemically resistant boots. A larger variety of sizes is now available, with more to come. Because of the limited number of polymers currently used, however, foot protection must be selected with care, since boots may act as 'chemical sponges.'

Partial body protection is normally provided by an apron, coverall, or rain-suit made from chemically resistant materials. A variety of construction materials is available along with chemical permeation performance data. The need for careful quality control of these materials is illustrated by the imperfections found in a butyl-coated fabric. Because they provide leak points for the CPC, clothing closures such as zippers, snaps, and flaps must be examined carefully.

Complete body protection is provided by TECP suits that extend the coverall or rain-suit design to a completely impermeable barrier surrounding the worker. All of the other major types of CPC are incorporated into a TECP suit design, plus specialized items such as vent valves, airline connections, body-cooling devices, communication equipment, and respirators. Special performance testing is required in the development and use of TECP suits and to a lesser extent for other types of CPC.

The informed health and safety professional can improve the CPC protection provided to the worker by considering clothing type, method of construction, and specific application when selecting chemical protective clothing for use in the workplace.

References

1. E.E. Havenor, *Test Report, Hypergolic Propellant Exposure Test of Candidate Propellant Handlers Ensemble Visors for Kennedy Space Center*, National Aeronautics and Space Administration, Washington DC, NASA TR-315-001 (1982).

2. James Burnett, *A.J. Chemical Accident, Benicia, CA*, National Transportation Safety Board, Washington, DC (1984).

3. National Fire Protection Association, *Open-Circuit Self-Contained Breathing Apparatus for Fire Fighters*, National Fire Protection Association, Quincy, MA (1987).

4. R.L. Mickelson and R.C. Hall, "A Breakthrough Time Comparison of Nitrile and Neoprene Glove Materials Produced by Different Glove Manufacturers," *Am. Ind. Hyg. Assoc. J.* **48**, 941 (1987).

5. American Society for Testing and Materials, "Standard Practice for Measurement of Dimensions," ASTM D3767, American Society for Testing and Materials, Philadelphia, PA (1979).

6. M.G. Olsen, "Specification for Environmental Control System, Propellent Handlers' Ensemble," National Aeronautics and Space Administration, J.F. Kennedy Space Center, FL, 79K20408 (1979).

7. M.G. Olsen, "Specification for Outfit, Protective Propellent Handlers' Ensemble," National Aeronautics and Space Administration, J.F. Kennedy Space Center, FL, 79K20409 (1979).

8. K. Ahmil and C. Kull, "Design and Qualification of NASA's Propellant Handlers' Ensemble" *Performance of Protective Clothing and Materials*, Second Symposium, ASTM STP 989, S.Z. Mansdorf, R. Sager, and A.P. Nielsen, Eds., American Society for Testing and Materials, Philadelphia, PA, 1988.

9. J.S. Johnson and J. Stull, "Measuring the Integrity of Totally Encapsulating Chemical Protective Suits," in *Performance of Protective Clothing and Materials*, Second Symposium, ASTM STP 989, S.Z. Mansdorf, R. Sager, and A.P. Nielsen, Eds., American Society for Testing and Materials, Philadelphia, PA, 1988.

10. J.O. Stull, J.S. Johnson, and P.M. Swearengen, "Hydrogen Fluoride Exposure Testing of the US Coast Guard's Totally Encapsulating Chemical Protective Suit," in *Performance of Protective Clothing and Materials*, Second Symposium, ASTM STP 989, S.Z. Mansdorf, R. Sager, and A.P. Nielsen, Eds., American Society for Testing and Materials, Philadelphia, PA, 1988.

11. American Society for Testing and Materials, "Standard Practice for Pressure Testing of Gas-Tight Totally Encapsulating Chemical Protective Suits," ASTM F1052, American Society for Testing and Materials, Philadelphia, PA (1987).

12. American Society for Testing and Materials, "Standard Practice for Qualitatively Evaluating the Comfort, Fit, Function, and Integrity of Chemical Protective Suit Ensembles," ASTM F1154, American Society for Testing and Materials, Philadelphia, PA (1988).

Additional Reading

J. Agranff, *Modern Plastics Encyclopedia* (McGraw Hill, Hightstown, NJ, 1985).

L.R. Birkner, *Respirator Protection, A Manual and Guideline,* American Industrial Hygiene Association, Akron, OH (1980).

A.D. Schwope, P.P. Costas, J.O. Jackson, J.O. Stull, and D.J. Weitzman, "Guidelines for the Selection of Chemical Protective Clothing," 3rd ed., (1987).

"Tingley Protective Clothing Brochure," Tingley, Inc., South Plainfield, NJ (1985).

Chapter 7

Personal Protective Equipment Selection Criteria and Field Use

James S. Johnson

Introduction

The preceding chapters have provided information on why the skin cannot provide protection from many toxic chemicals, a brief description of polymer chemistry and permeation through polymers, a review of chemical protective clothing test methods, and a description of various CPC construction techniques. This chapter will describe how to select and use chemical protective clothing.

From the outset, health and safety professionals must understand that many CPC selections may not be clearcut choices. Quantitative data for CPC performance has been available only since about 1977. Most of this information characterizes the individual chemical resistance of new barrier materials. Data about the chemical resistance of seams, closures, and other CPC components may or may not be available.

With these limitations one must wonder how CPC can be selected properly and without any risk to the user. By careful evaluation of the hazards requiring protection, a selection can be made that *minimizes risk* to the user but does not completely eliminate it. This reality of CPC use creates frustration in many health and safety professionals who expect their recommendations to be risk free. Because the field is still quite young, this shortcoming must be understood, accepted, and taken into consideration when selecting CPC. Before making any CPC selection, it is imperative that the health and safety professional visit or be intimately familiar with the process under consideration to make sure the equipment selected is performing adequately. Routine followup evaluations should also be carried out to assure CPC performance on a long-term basis.

Preliminary Workplace Survey

The first step in the CPC selection process is an evaluation of the workplace and the specific process that requires CPC. As part of this evaluation, the health and safety professional should always be looking for ways to use engineering controls or the substitution of less-toxic materials to remove the need for CPC in the first place.

Once the need for CPC is established, one must gather specific details for use in the selection process. This evaluation should always include experience from similar applications in other industries. If the process under evaluation is using chemicals purchased from another source, the chemical manufacturer is a good contact point to identify the type of CPC used in their own operations.

In Chapter 9 of this textbook a CPC selection worksheet has been developed to aid in the evaluation. Key initial-evaluation information contained on that sheet is listed below.

Job Classification or Task

- Identify/characterize the task accurately.

Process or Task Summary

- Summarize the operation, providing specific details.
- Determine type of operations where CPC will be used (e.g., dipping, spraying, mixing, moving of components or parts, assembly or disassembly work, repair, demolition, emergency response).
- Note any other tasks that should be evaluated because they impact the CPC performance of the task at hand.
- Note the normal variation in the workplace from summer to winter, day to night.

Potential or Actual Chemical Hazards

- Identify specific chemicals involved in the task.
- For overlapping tasks, identify the chemicals and their effects.
- Note the temperature of chemicals.

Physical Properties of Chemicals

- Identify as a solid, liquid, or gas.
- List the material's vapor pressure.
- If mixture, list properties of major components.

Potential or Actual Physical Hazards

- Note physical hazards (e.g., heat stress, electrical shock, head injury, slip/trip hazards, foot hazards, cut, puncture, abrasion risks).

Chemical Contact Periods

- Determine if contact with the chemical(s) will be as solids, liquids, vapors, or gas.
- Note how long the CPC will be in direct contact with the chemicals (seconds, minutes, hours, days [reuse?]).
- Determine if the contact is routine, intermittent, or infrequent/unplanned.
- Note if the CPC is simply for splash protection, and if the wearer can change quickly.
- Note if the CPC will be used for emergency response.

Type of Potential Contact

- Determine the type of chemical contact expected from routine and emergency conditions (e.g., routine splash, pressurized spray accident).

Body Zones of Potential Contact

- Refer to Chapter 6 (Figure 6-1 and Table 6-1) of this textbook for a summary of the types of CPC equipment to consider.

If a good Material Safety Data Sheet is available, it can be attached to this analysis and pertinent sections referred to. With this information identified, the health and safety professional is now ready to start the CPC evaluation/selection process.

Chemical Toxicity

A very important parameter that must be carefully determined is the toxicity of the chemical against which the CPC will provide protection. This is not a straightforward and simple task, however, because no published standards exist for skin exposure to toxic chemicals.

The American Conference of Governmental Industrial Hygienists (ACGIH) identifies chemicals with a "Skin" notation in their *Threshold Limit Values and Biological Exposure Indices* booklet.[1] The "Skin" notation refers to a potential contribution to the overall exposure by the cutaneous route (including mucous membranes and eye), either by airborne exposure or by direct contact with the substance. A review of the consistency that exists between chemicals with a "Skin" notation found large variations in the toxic properties of these chemicals and their effects on humans.[2]

A statistical technique of multiple linear regression and discriminant analysis has been used to determine if certain chemical/physical properties can be used to identify chemicals that should have a Skin notation.[3] A series of 62 industrial organic compounds bearing "Skin" notations was chosen and compared with 23 organic compounds without such a notation. The study found no significant correlation among Hildebrand solubility parameter, molecular weight, melting point, dielectric constant, and documented skin toxicity. A statistically significant correlation was found for water solubility and vapor pressure, however, and a more significant correlation was found based on toxicity alone, with such indicators as Threshold Limit Value (TLV), the lethal dose that kills 50% of the exposed test population (LD_{50}), and the Approximate Lethal Dose (ALD).

Another way to express skin hazard is[4]:

Skin Hazard $\propto$ (Toxicity × Bioavailability).

The evaluation process must therefore use information on the toxicity and bioavailability of the chemical(s) in question. If acute oral toxicity data is used, certain derived classifications and rankings are available, but these should be used only as initial screening tools. Table 7-1 lists a toxicity ranking chart for chemicals based on LD_{50} and LC_{50} (lethal concentration in air that will kill 50% of an exposed rat population).[5] Table 7-2 lists the LD_{50} of some common toxic materials to provide an example that uses actual acute toxicity data; a skin absorption estimate technique based on animal toxicology data is also available for certain applications.[6] Table 7-3 extends the estimation process to humans, but remember that all of these tables are only qualitative at best and must therefore be used carefully.[7]

At the present time, we have no simple method to summarize bioavailability, so this part of the evaluation must take into account such things as water solubility and specific metabolism routes to provide an estimate of bioavailability.

As part of a proposed dermal hazard rating scheme, Eiser has identified a number of factors that should be considered in relation to effects on the worker or the test animal.[8] The chemicals analyzed in this dermal scheme were taken from the ACGIH TLV Booklet 1987–88. These factors are listed below by categories and should be considered in the evaluation of the toxicity of chemicals for which CPC will be used.

"Skin" Designation Chemicals

- Potential for systemic effects due to skin absorption alone.
- Anticipated severity of systemic effects.
- Reversible vs permanent damage.
- Speed of action on the body.
- Warning properties.
- Absorption rate (skin permeation rate) for the substance.
- Known significant toxic dose.
- IDLH (Immediately Dangerous to Life or Health) potential with skin contact alone.
- Chronic hazard potential, such as carcinogenic or teratogenic effects.
- Reported fatalities with use.

Corrosives or Irritants (Not Assigned a "Skin" Notation)

- Necrotic skin effect vs transient irritation.
- Local and systemic effects vs local effects alone.
- Draize scores, related refinements, and alternate test data.
- Warning properties.
- Permanent vs temporary injury or disability.
- History of industrial or laboratory use.

Allergic Sensitizers

- Potent vs mild sensitizers.
- Anaphylactic shock responses reported with use.
- Acute toxic or corrosive effects with skin contact in addition to sensitizing potential.
- Chronic toxic effects in addition to sensitizing potential.

Table 7-1. Toxicity classes.

Toxicity rating	Descriptive term	LD_{50}–wt/kg single oral dose (rats)	4-h inhalation LC_{50}–ppm (rats)
1	Relatively harmless	15 g or more	>100,000
2	Practically non-toxic	5–15 g	10,000–100,000
3	Slightly toxic	0.5–5 g	1000–10,000
4	Moderately toxic	50–500 mg	100–1000
5	Highly toxic	1–50 mg	10–100
6	Extremely toxic	1 mg or less	<10

Table 7-2. Approximate acute ingestion LD_{50}s of a selected variety of chemical agents.

Agent	LD_{50} (mg/kg)
Ethyl alcohol	10,000
Sodium chloride	4000
Ferrous sulfate	1500
Morphine sulfate	900
Phenobarbital sodium	150
DDT	100
Picrotoxin	5
Strychnine sulfate	2
Nicotine	1
d-Tubocurarine	0.5
Hemicholinium-3	0.2
Tetrodoxin	0.10
Dioxin (TCDD)	0.001
Botulinus toxin	0.00001

Table 7-3. Toxicity rating chart.

	Probable oral lethal dose for humans	
Toxicity rating or class	Dose	For average adult
1. Practically non-toxic	>15 g/kg	More than 1 quart
2. Slightly toxic	5–15 g/kg	Between pint and quart
3. Moderately toxic	0.5–5 g/kg	Between ounce and pint
4. Very toxic	50–500 mg/kg	Between teaspoonful and ounce
5. Extremely toxic	5–50 mg/kg	Between 7 drops and teaspoonful
6. Supertoxic	<5 mg/kg	A taste (less than 7 drops)

CPC Performance Information

Degradation Data

Chapter 5 described in detail the degradation, penetration, and permeation test methods used for candidate CPC materials. To date, not one accepted test method is recommended to characterize CPC degradation. Degradation data, however, provides an assessment of gross changes in the CPC construction material and can be useful for coarse screening of candidate clothing products. The experimental results are summarized in general terms, such as "recommended," "minor effects," "moderate effects," and "not recommended." This information can also be useful when evaluating used CPC to identify gross changes that may have been missed as part of the initial selection process. Stampfer et al. reported the 24-hr weight and volume changes for ten different CPC garment materials which were exposed to four chemicals.[9] Table 7-4 presents weight and volume change results from this study. It is obvious from these results that the CPC garment materials are being affected by the 24-hr chemical exposure. From this data, candidate materials can be chosen for additional penetration and permeation studies. Crude CPC selections could be made from this type of data, but only with great care.

Table 7-4. Twenty-four hour weight and volume changes.

Material	Epichlorohydrin Weight %	Epichlorohydrin Volume %	Perchloroethylene Weight %	Perchloroethylene Volume %	Trichloroethylene Weight %	Trichloroethylene Volume %	1,2-Dibromoethane Weight %	1,2-Dibromoethane Volume %
Surgical Rubber	30	30	770	530	700	580	480	240
Butyl Rubber	3	0	510	280	440	320	65	30
Polyethylene	0	15	15	85	20	70	20	35
Polyvinyl Alchohol	-7	-7	-6	-12	-2	-10	4	0
Neoprene	100	120	360	320	400	410	500	[a]
Nitrile	340	240	95	60	310	220	580	230
Viton	20	35	4	0	8	20	3	0
Viton SF	25	35	4	0	12	15	3	0
Vitrile	480	350	65	45	290	200	690	290
Teflon	0	0	0	0	0	0	2	0

[a] Sample distintegrated.

The importance of color change, stiffening, swelling, tackiness, etc. of the CPC construction materials must be emphasized to the user because these symptoms may indicate significant deterioration and loss of protection. These changes should also be reported to the CPC program administrator so that a proper assessment can be made.

Penetration Data

Penetration data is available in a more consistent and comparable form because an ASTM test method (F903) is available. This test provides a simple pass/fail criterion that permits easy selection of acceptable CPC items. This test, though simple, is important in determining the resistance to chemical flow provided by a candidate barrier material, seam, or closure. Since the test is carried out for ten minutes under a pressure of 1 psig, a reasonable measure of general barrier resistance of the CPC is obtained. Evaluating the penetration data for a variety of chemicals against which the CPC will protect can result in important information for use in the final selection process.

Permeation Data

Permeation data, developed using the ASTM Standard Test Method F739, constitutes the largest information base currently available to make CPC selections. Two forms of data are available from this test method—breakthrough time (BT), and steady-state permeation rate (SSPR). Breakthrough time is the time at which the test chemical is first detected on the clean side of the CPC barrier; the steady-state permeation rate is a measure of the amount of hazardous chemical that will pass through the barrier per unit time. It must be remembered that the F739 method only tests a piece of the CPC item, the entire item is not tested. The potential for permeation through discontinuities is not determined using this test method. Because there are no quantitative standards for skin exposure to hazardous chemicals (which would permit the use of SSPR data), more emphasis for CPC selection has to date been placed on the breakthrough time.

When making CPC selections based on BT, it must be remembered that the sensitivity of the analytical detection method can significantly affect the measured BT. The need to standardize the detection limit and to calibrate the complete system is presently being considered in a revision of the ASTM F739 test method. To illustrate how significant differences in detection limit can affect the BT, Table 7-5 compares the BT of Teflon-coated Nomex vs acetone for two different analytical methods.[10]

Table 7-5. Breakthrough time of Teflon-coated Nomex vs acetone.

Detection method	Breakthrough time
^{14}C-labeled acetone scintillation counting	45 min
Gas chromatography flame ionization detection	>240 min

With radioactive-labeling detection techniques, one can obtain high sensitivity levels.[11] In the case of Viton/chlorobutyl vs ^{14}C-labeled acetone, breakthrough is seen almost immediately (<2 minutes), where gas chromatography flame ionization detection found a breakthrough time of 54 minutes. Other reported BTs for this material range from 43–98 minutes. It can be seen that large differences in BT are possible if different chemical detection systems are used. If the calibration process looks only at the detector and not at the complete sampling system, large differences in BT are also possible.

The use of radioactive tracers indicates another possibility that must be considered—What is the exposure level to the wearer even before the reported breakthrough time? Can it be assumed to be zero? Because some chemicals at a molecular level can permeate through organic barrier materials very quickly (seconds), extremely low concentrations of permeating chemicals are possible long before the analytical techniques can detect them. This is a reality anyone selecting CPC must understand. Other performance techniques such as biological monitoring and air sampling must therefore be used to completely evaluate CPC performance. This information should also be presented to the users of CPC in a non-alarming manner, when appropriate, so that the limitations of the analytical test method and the CPC garment are understood.

In a recent study Harville and QueHee report on the permeation of a 2,4-D isooctyl ester formulation through neoprene, nitrile, and Tyvek protective materials.[12] The reported detection limit for the 2,4-D isooctyl ester was quite sensitive, 115 pg. The results of three nitrile glove permeation graphs are shown in Figure 7-1. By examining the x-axis it can be seen that a zero breakthrough occurs with varying concentration levels. The difference between the formulation liquid permeation and an aqueous mixture is also shown. The criteria used for choosing the best glove from the permeation data were the following: time elapsed between initial contact and no more than 10 μg transferred; rate of any permeation over this period; amount detected initially; amount detected in the 2-hr period, and the amount detected over 8 hr. From these analyses Tyvek and a specific brand of nitrile gloves were chosen. The authors noted that their analysis may be unacceptable if the 2,4-D isooctyl ester were a carcinogen.

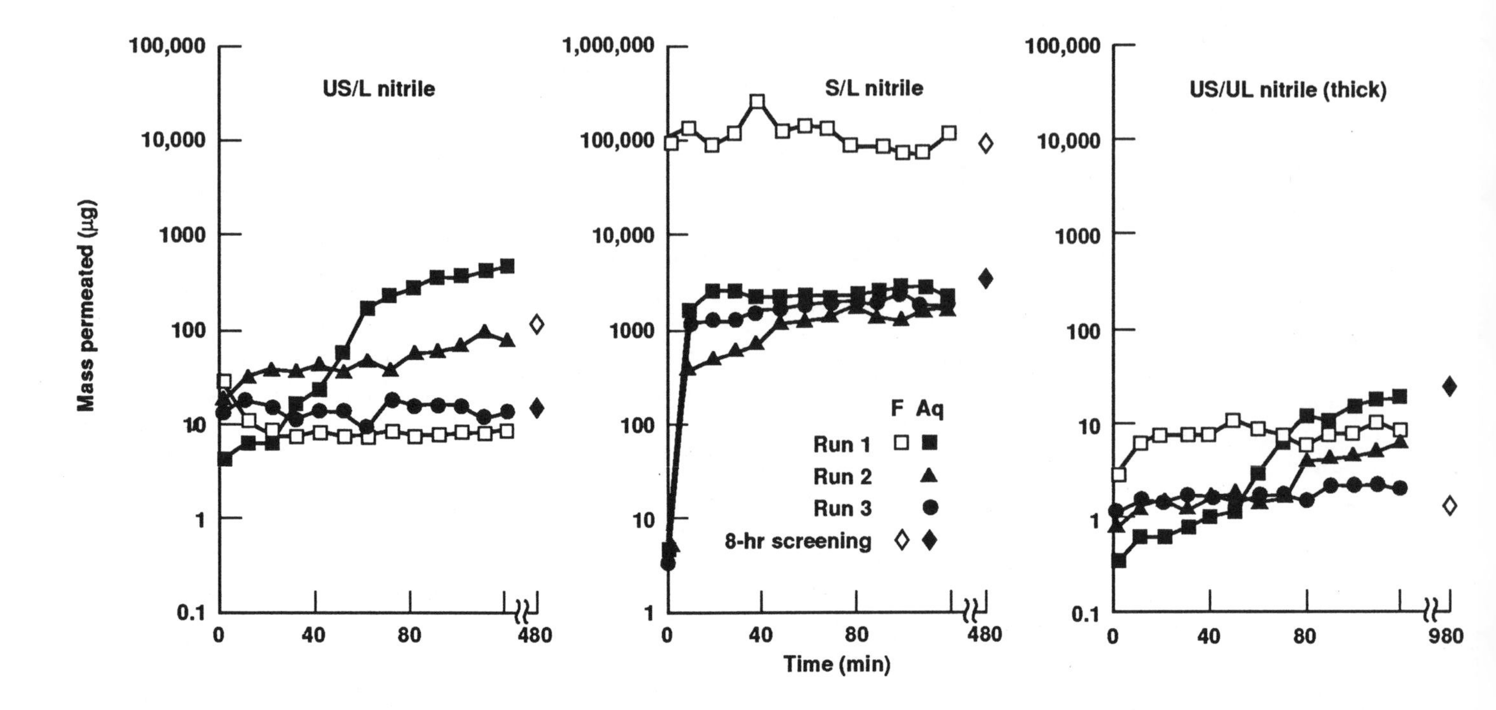

Figure 7-1. Permeation/penetration behavior of 2,4-D isooctyl ester through unsupported/lined (US/L), supported/lined (S/L), and unsupported/unlined (US/UL) glove materials. Note that even at Time 0 some permeation was reported in every instance.

Steady-state permeation rate data can be used to aid in the selection of the best CPC for a specific application. If the SSPR is used as a measure of the amount of exposure a CPC wearer can experience after breakthrough, useful barrier performance evaluations can be made. Nelson et al. have identified five different permeation rate vs time behaviors, which are shown in Figure 7-2.[12]

Type A, the most typical of the five types of permeation behavior, occurs when the specimen remains physically unchanged during the test. The glove offers complete protection until initial breakthrough is observed at point 1. The permeation rate continues to increase in the characteristic "S"-shaped curve until a maximum rate, point M, is obtained; the rate then stabilizes at this equilibrium value.

Type B behavior is essentially the same as Type A, except that after point M is reached, the permeation rate slowly increases or decreases. This phenomenon occurs when the sample is structurally modified by the solvent, either by chemical reaction or some physical deformation.

Type C behavior occurs when the glove exhibits a catastrophic breakthrough. The glove behaves normally until point X is reached, at which time a full-scale analyzer response is observed as the sample dissolves, admitting the entire volume of test solvent into the overflow trap.

Type D response happens when the maximum permeation rate is normally achieved; the rate decreases sharply and attains an equilibrium value at point Y. This type of behavior occurs when the rate eventually stabilizes at a time near point Y.

The double "S" shape of Type E shows an inflection at point Z that occurs when the sample undergoes a high degree of swelling. Although a similar pattern was observed by Weeks and Dean,[13] no satisfactory explanation has been postulated.

If SSPR data is available and used as part of the CPC selection process, a complete curve for the use time in question should be evaluated. Figure 7-3 shows plots of two simplified extremes in SSPR. It is obvious that Barrier 2 fails almost catastrophically, resulting in a large potential exposure to the user. In the case of Barrier 1, the exposure potential is low and the user has time to take action. The selection process should stay away from Barrier 2 SSPR performance and choose materials with behavior as

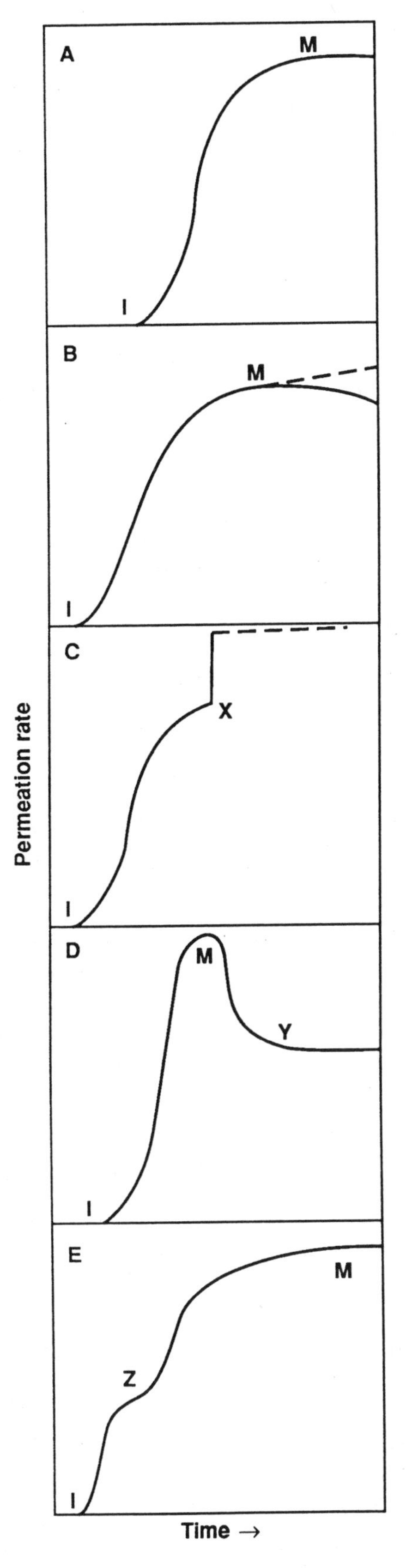

Figure 7-2. Five types of permeation behavior.

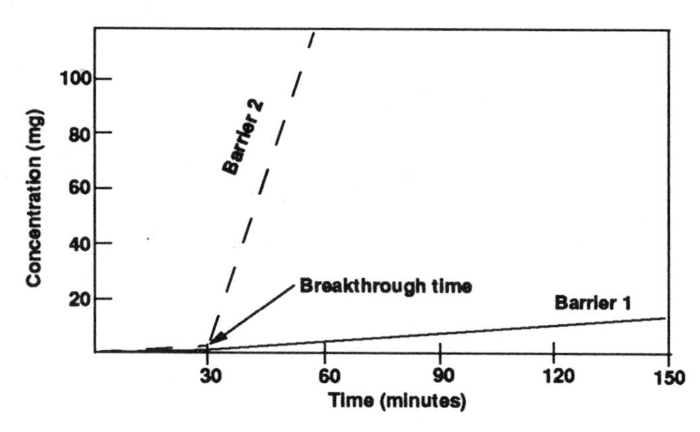

Figure 7-3. Plots of the two extremes in steady-state permeation rate vs time.

close to Barrier 1 as possible, especially when highly toxic chemicals are of concern.

Temperature Effects

Chemical

The effects of temperature on CPC performance can be significant and should be considered in the selection process. In most cases, the BT and SSPR are measured at room temperature 70°F (25°C), while actual use temperatures can be expected to vary depending on the workplace conditions. Remember also that body temperature is 98.6°F (37°C), which will become the temperature of the inner layer of CPC. In Table 7-6 the effects on BT due to increasing the temperature are listed for dichloromethane and methyl ethyl ketone against Viton/chlorobutyl laminate.[15]

A temperature change from 41°F (5°C) to 113°F (45°C) results in a BT decrease from 8–10 minutes to instantaneous in one case and a decrease by a factor of 13 in the other case. These temperature fluctuations, causing the BT effects, are not extreme and represent ordinary summer and mild winter temperatures during a typical work year, for example, or going from the inside to outside in the winter. Table 7-7 lists the smaller changes in BT for two typical barrier materials against acetone.

A significant reduction is seen in BT for Viton/chlorobutyl laminate, by almost one half for a ΔT of 12°F (6.5°C). For the 28-mil chlorinated polyethylene, a ΔT of only 4°F (2.5°C) results in an 11–15% decrease in BT. Table 7-8 lists the effect of temperature variation on BT and SSPR for neoprene against benzene and Viton against methylene chloride. Forsberg et al. also observed a 55% decrease in BT for benzene penetration through neoprene for a ΔT of 86°F (30°C), and Perkins reports a decrease of 43% in BT for a ΔT of 50°F (10°C).[16,17]

Perkins also notes that the temperature effect on permeation parameters can be determined by the Arrhenius relationship if permeation data is available for two or more temperatures. This is not normally the case, but if experiments are planned, multiple temperature evaluations should be included in the experimental protocol. In general, CPC users should remember the following with reference to temperature changes: Reduced temperature results in a reduced SSPR and an increase in BT. An increased temperature results in an increase in SSPR and a decrease in BT.

Table 7-6. The effects of temperature on dichloromethane and methyl ethyl ketone breakthrough times for Viton/chlorobutyl laminate.

Temperature °F (°C)	Breakthrough times (min)[a]	
	Dichloromethane	Methyl ethyl ketone
41 (5)	8–10	180–199
59 (15)	6–8	80–85
77 (25)	4–4.5	35–45
95 (35)	3.5–4	20–25
113 (45)	—	14–20

[a] Breakthrough times measured using ASTM F739.85; dichloromethane breakthrough measured with GC with ECD; MEK measured by GC with FID.

Table 7-7. The effect of ambient temperature on BT of acetone vs Viton/chlorobutyl laminate and chlorinated polyethylene.

Test material	Temperature °F (°C)	Acetone BT (min)
Viton/chlorobutyl laminate	68 (20)	95–98
	80 (26.5)	43–53
28-mil chlorinated polyethylene	72 (22)	32–35
	76 (24.5)	27–31

Table 7-8. The effect of temperature on BT and SSPR for neoprene against benzene and Viton against methylene chloride.

Temperature (°C)	Breakthrough time (min)	Steady-state permeation rate (mg/cm^2-min)
Benzene permeation through neoprene[a]		
7	40	0.13
22	24	0.16
37	16	0.23
Methylene chloride permeation through Viton[b]		
25	61	0.04
30	47	0.054
35	35	0.067

[a] From Forsberg et al.[16]
[b] From J. Perkins.[17]

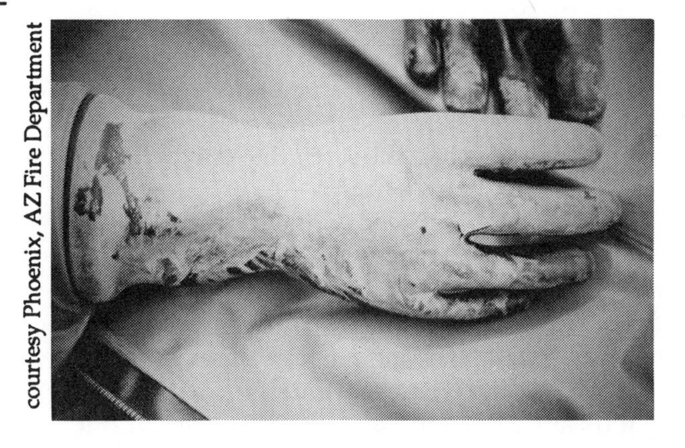

courtesy Phoenix, AZ Fire Department

Figure 7-4. The neoprene glove has cracked from exposure to liquid chlorine.

Physical

The effect of temperature on the physical characteristics of polymers is well known. As the temperature rises above RT, most polymers become softer and more pliable. At some temperature, however, the effects will be deleterious to the strength properties of the polymer. As the temperature drops below room temperature, the polymer becomes stiffer and reaches a point where it becomes brittle and will crack and break. See Chapter 3, Polymer Transitions, for more details. In field selection, it must be remembered that temperatures will affect the physical properties of the CPC selected, e.g., Saranex becomes brittle at –20°F and cracks. It has also been reported that several emergency response workers in TECP suits knelt in liquid chlorine, stood up and the knee areas of their suits cracked. Figure 7-4 illustrates a neoprene glove failure due to the cracking effect of liquid chlorine.

Pressure Effects

The effect of pressure on BT and SSPR has not been studied in detail, nor are its effects considered in the currently used permeation test method. Gunderson et al. presented an interesting study on the selection of gloves for protection against an amine hardener composed of 57% meta-phenylenediamine (MPDA), 30% methylenedianiline (MDA), and 13% diglycidyl-ether of bisphenol A (DGEBA).[18] The characteristic of this hardener is that even minute amounts of MPDA produce a yellow-brown stain that is extremely difficult to clean from surfaces and permanently discolors the skin. Visual observations of glove failure became simple and obvious. A series of modified permeation tests were carried out as well as several ASTM F739 tests to identify nitrile gloves as good candidates for use against this amine hardener.

To evaluate the effectiveness of these gloves in actual workplace applications, a unique detection system was developed. New white cotton gloves were worn under the nitrile gloves as the "detector." The first observation noted is that the type of work carried out can affect MPDA permeation. A 20-minute disassembly operation (where the worker used nitrile gloves) resulted in significant staining of the internal glove, whereas the nitrile glove laboratory test data indicated a breakthrough time of over one week. Similar results were observed for several workers carrying out the same task using the same type of gloves. The stains observed on the white detector gloves were on the contact/pressure areas where the hand touched the screwdriver handle.

After a glove had been contaminated with the amine hardener from a short-duration operation (15 minutes), it was obvious the next day by color that the entire glove had become contaminated. This observation resulted in single-use requirements for the gloves. This very practical and informative study took advantage of the visible detection of trace quantities of a chemical to show the potential shortcomings of glove selection based on laboratory data. Normal pressure accompanying the use of simple hand tools decreased the breakthrough time and possibly increased the SSPR significantly. Reuse of gloves contaminated the previous day was not acceptable because of visible staining.

This amine hardener's effect on gloves was easy to evaluate because of the visible residue, but this is not the usual situation. Care must therefore be exercised with other chemicals to make sure no significant increases in permeation, or decreases in BT, result from actual use. Evaluation of penetration data must also be done with reference to the effects pressure may have on the barrier's performance.

Personal monitoring and CPC evaluations are necessary on a routine basis to accomplish this. Forsberg evaluated the amount of internal fluorescent particulate contamination that results from reuse by looking at the outsides and insides of several pairs of gloves under ultraviolet light (Figure 7-5).[19] From these examples, it is obvious that chemical exposure can result from glove reuse (which is dependent on brand and type), exposure environment, and employee work practices. The question of reuse and decontamination followed by reuse must be analyzed carefully to assure no toxic exposure to the CPC user. See Chapter 8 for a more detailed discussion of decontamination.

Eggestad and Johnson have recently reported the results of experiments carried out to evaluate the effects of liquids under pressure on the penetration and wicking of repellent fabrics.[20] The permeable adsorptive fabrics' penetration was significantly affected by a pressure of 2.0 kg/cm^2 (Figure 7-6) whereas the wicking fabric was less affected (Figure 7-7). This study indicates that penetration can be affected by pressure, thus requiring a careful analysis of pressure effects when penetration data is used in the selection process for CPC.

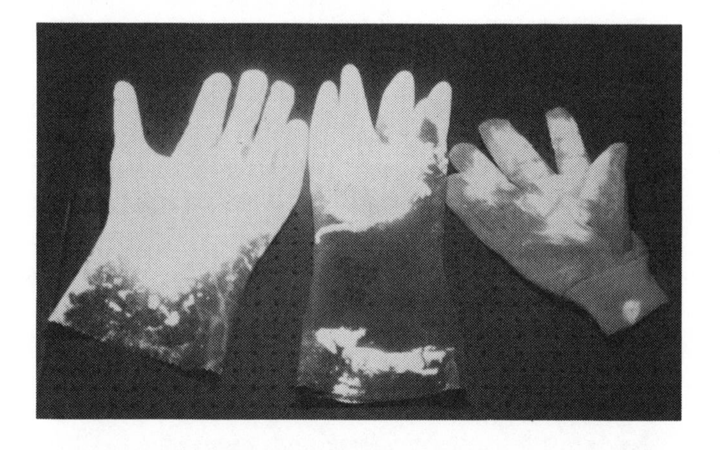

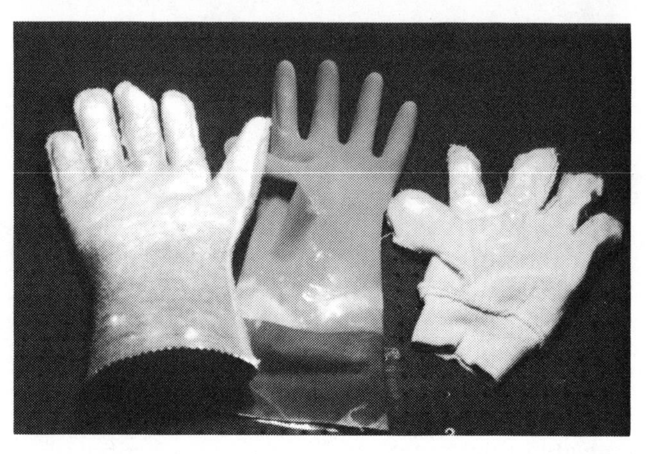

Figure 7-5. Contamination visible on the outsides (top) and insides of gloves under ultraviolet light.

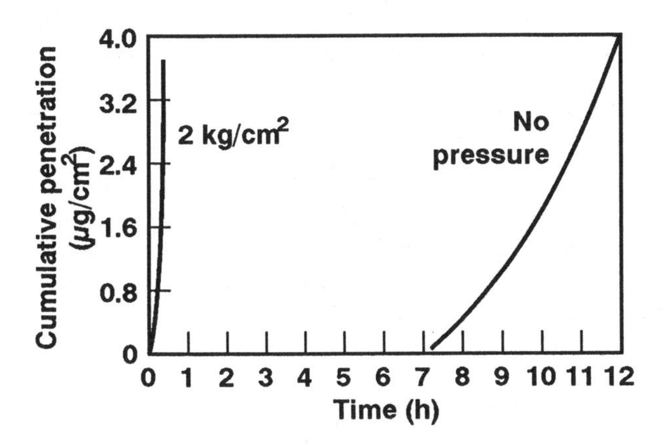

Figure 7-6. Cumulative penetration vs time for specimen with repellant cover fabric. Note the effect of pressure on the breakthrough time.

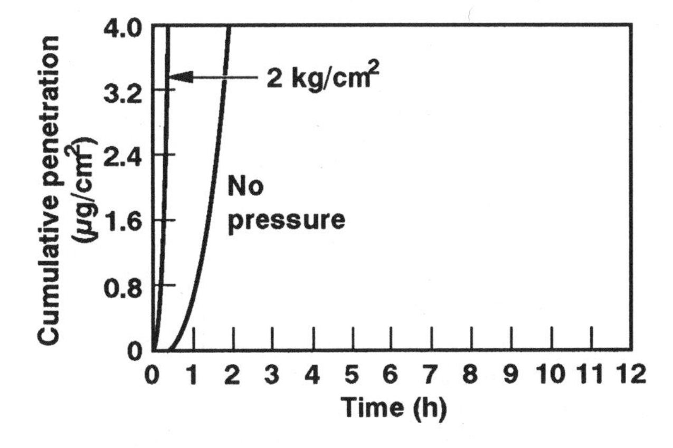

Figure 7-7. Cumulative penetration vs time for specimen with wicking cover fabric. Pressure effects are still apparent, but not as great.

Barrier Thickness Effects

The thickness of the glove material is another parameter that must be evaluated carefully. The wearers will generally be interested in a thinner glove, which offers greater dexterity. Nelson et al. evaluated the effect of glove thickness on permeation rate for four different glove/solvent systems. These results are presented in Figure 7-8. All four glove/solvent pairs demonstrated the permeation rate to be inversely proportional to the thickness, which is in agreement with Fickian diffusion (described in Chapter 4).

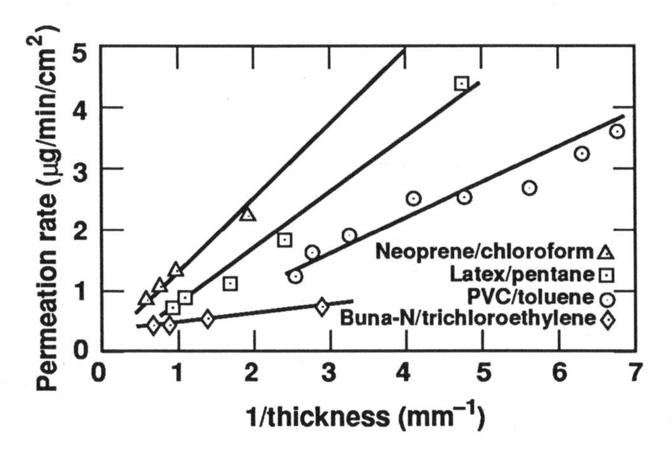

Figure 7-8. Effect of glove thickness on permeation rate.

Perkins has noted that when diffusion is non-Fickian, permeation rates may not vary with the thickness because an almost-constant front of the challenge chemical advances across the protective barrier material. In these cases, breakthrough times will increase with thickness, while the permeation rates will not change.

Nelson et al. also noted that care must be used when generally applying this inverse relationship of thickness to permeation rate of different brands of gloves made from the same polymer. Experimental data for different brands of neoprene gloves produced inconsistent permeation rates, though the general material used—neoprene—was the same. Schlatter and Miller, in a series of studies using hexane and methyl isobutyl ketone against butyl, neoprene, and nitrile glove polymers, found a 25% and 95% deviation in BT from the inverse proportionality relationship.[21] They also noted that increasing thickness generally increased breakthrough time by a value larger than the proportional amount. In addition, they noted two cases where the BT increase was less than the relative thickness increase.

With this information in mind, estimating breakthrough change based on changes in thickness without experimental data should be done with care. It is reasonable to use a linear relationship between BT and thickness with the understanding that many cases will include a good safety factor, but in a few cases there will be no safety factor—and the breakthrough time could actually decrease.

Chemical Mixture Effects

In the actual work environment, most CPC exposures occur with chemical mixtures. The obvious tendency is to evaluate a mixture based on the BT and SSPR of its components. Nelson et al. looked at the effect of mixtures on CPC with a variable chloroform/toluene mixture against neoprene, and a pentane/trichloroethylene mixture against polyethylene. In the case of the neoprene polymer, no unusual results were observed; but in the case of polyethylene, they found a large positive synergistic effect on the 50/50 permeation rate. The authors used these results to conclude that permeation rates of mixtures cannot be accurately predicted if the rate of only one component is known.

The need to base breakthrough times of mixtures on actual data is reinforced by the CPC performance data reported by Stull and listed in Tables 7-9 and 7-10. Table 7-9 examines the BT of dichloromethane (CH_2Cl_2) in hexane and toluene against Viton/chlorobutyl laminate.

In both systems the second solvent (hexane or toluene) does not break through the laminate in its pure form, whereas CH_2Cl_2 has a breakthrough time of 25–36 minutes. For a 50/50 mixture (by volume) of either CH_2Cl_2 and hexane or CH_2Cl_2 and toluene, both mixture components do permeate the material samples. For both mixtures, CH_2Cl_2 initially broke through at a time somewhat longer than its normal BT for the laminate, with the second solvent permeating about ten minutes later. It is suspected that the CH_2Cl_2, which readily permeates the Viton/chlorobutyl laminate, carries the second solvent through. This conclusion was also reached in other mixture studies carried out by Forsberg et al. and Mickelson et al.[22,23] Mickelson et al. noted that since the permeation of binary mixtures through CPC cannot be predicted by the permeation results of the pure component, they recommended that CPC be tested for its permeation characteristics with the use of the chemical mixture concentrations and conditions that reflect the work site exposures.

Table 7-10 presents the permeation breakthrough time for acetone/hexane mixtures against Viton/chlorobutyl laminate.

Acetone has a normal breakthrough time of 53–61 minutes, whereas hexane does not permeate the Viton/chlorobutyl laminate within three hours. However, any combination of hexane and acetone results in a significantly shorter breakthrough time. In fact, breakthrough time occurs within ten minutes of initial mixture contact with the laminate in many cases. Furthermore, both acetone and hexane break through the laminate simultaneously, as detected by gas chromatography. This synergistic effect of the two chemicals cannot be explained in terms of individual effects on the material by the two chemicals.

The limited data summarized in this section on the permeation of mixtures indicates that conclusions based on only one of the components must be done very cautiously. Ideally, laboratory experimentation should substantiate any conclusion based on single-component permeation data.

Table 7-9. Permeation breakthrough times for two binary mixtures against Viton/chlorobutyl laminate.

		Breakthrough time (min)	
% CH_2Cl_2	No. runs	CH_2Cl_2	Hexane
In hexane:			
100	1	25–36	—
50	2	42–47	57–62
0 (100% hexane)	4	—	no BT
In toluene:			
100	1	25–36	—
50	1	45–55	58–66
0 (100% toluene)	1	—	no BT

Table 7-10. Permeation breakthrough times for acetone/hexane mixtures against Viton/chlorobutyl laminate.

Percentage acetone	Number runs	Breakthrough[a] time (min)
100	7	53–61
95	1	0–5
86	1	6–11
50	5	2–6
35	2	0–6
15	1	6–11
5	1	0–5
1	1	0–5
0 (100% Hexane)	4	no BT (3 hrs.)

[a] Breakthrough times reported for both acetone and hexane.

Manufacturer and Model-Specific Performance

It would be nice to be able to simplify the process of CPC selection to identification of the polymer by its generic type and thickness. This, unfortunately, is not possible due to variability in chemical compositions and fillers used to manufacture the various commercially-available polymers of the same generic type. Nelson et al. noted that neoprene glove polymers varied so much in permeation rates from brand to brand that only results from the same brand were compared in their report. Mickelson and Hall have carried out a study to evaluate the extent of difference in breakthrough times for nitrile and neoprene gloves against three chemicals.[24] The three test chemicals used to evaluate the nitrile glove polymers were perchloroethylene, p-xylene, and n-butyl acetate; for the neoprene glove polymers the chemicals were n-butyl acetate, n-hexane, and ethanol. The largest variation in BT was a factor of 10. The conclusion from this study noted that BT data for use in the selection of CPC or in predictive modeling should be manufacturer and product specific. Product information supplied by most major manufacturers of CPC now provides a good database for their products. If you have a specific permeation need, the CPC manufacturer's chemists may actually run the required test(s) for you, which is now looked upon as good technical support for regular customers. The data resulting from these tests generally can also be included in the next revision of their own permeation database.

Chemical Contact Time

V.I. Man et al. have recently reported on the effects of multiple splashes of various chemicals on the BT of Viton/chlorobutyl laminate and chlorinated polyethylene, as listed in Table 7-11.[25]

In this evaluation of BT, differences resulting from liquid contact, liquid splash (at three frequencies—12×, 6×, and 1×) and saturated vapor (at 0° and 25°C) have been made. In some cases BT increased with diminishing chemical contact (constant liquid contact >12× splash >6× splash >1× splash >25°C vapor and >0°C vapor); in other cases the material/chemical combination produced relatively constant BT with varying liquid contact times. BTs for these combinations were essentially independent of liquid chemical/material contact time. The observed variations were explained by differences in the test chemicals' ability to wet the surface of the barrier material. Additional evaluations of these exposure effects are required before physical parameters of the chemical and barrier material can be used to predict BT.

Berardinelli and Moyer have reported a significant difference in permeation parameters of respirator valve polymers, depending on whether the permeant test chemical is in the vapor or liquid form.[26] For methyl isocyanate a vapor concentration of 800 ppm MIC had a BT of >45 minutes, whereas the neat liquid was very permeable.

Table 7-11. The effects of multiple splashes of several chemicals on the BT of Viton/chlorobutyl laminate and chlorinated polyethylene.

Material/chemical combination	Breakthrough time (min.)					
	Liquid	Liquid splash[a]			Vapor	
		12×	6×	1×	25°C	0°C
Viton/chlorobutyl laminate:						
Acetone	43–58	43–58	73–78	94–100	63–74	3 hrs
Dichloromethane	25–36	30–35	30–35	30–35	35–55	3 hrs
Methyl ethyl ketone	25–40	35–40	35–40	50–55	80–85	3 hrs
Tetrahydrofuran	9–11	11–17	11–17	11–17	35–45	3 hrs
Chlorinated polyethylene:						
Acetone	32–35	50–53	68–72	75–85	130–140	3 hrs
Chloroform	30–37	46–50	81–86	120–125	132–138	[b]
Dichloromethane	15–24	20–26	25–30	26–32	32–40	3 hrs
Methyl ethyl ketone	28–35	40–45	45–50	46–49	141–148	3 hrs
Tetrahydrofuran	27–39	39–45	51–58	62–72	105–111	3 hrs

[a] Liquid splash testing: 12×—one splash every 15 minutes; 6×—one splash every 30 minutes; 1×—one splash at beginning of test.
[b] Test not performed.

Example of a CPC Selection Based on Laboratory Experimentation

A fine example of how an organization should carry out the CPC selection process was presented by the Syntex Corporation at the American Industrial Hygiene Conference in San Francisco, 1988.[27] The purpose of the study was to determine which gloves provide protection from enprostil in propylene carbonate and how long such protection would last. In production operations workers routinely handle solutions of propylene carbonate and enprostil to a maximum concentration of 6.72%. Enprostil is a type-E prostaglandin used for treatment of gastric and duodenal ulcers with the systemic effects of diarrhea, antifertility, platelet aggregation, and bronchioconstriction; however, it is a skin irritant. The need to protect workers from this chemical is obvious.

The investigators designed the evaluation to look for different gloves that would be acceptable for the job. The testing procedure followed the ASTM F739 method, and the chemical analysis used gas chromatography/flame ionization detection for propylene carbonate (1 mg sensitivity) and a radioimmunoassay method for enprostil (60 pg sensitivity). The experimenters recognized that their chemical of concern, enprostil, was dissolved in propylene carbonate, thus making the CPC-selection problem that of a chemical mixture. The permeation characteristics of propylene carbonate were first evaluated, followed by a mixture of 6.72% enprostil in propylene carbonate. Table 7-12 lists the results from this evaluation.

The data indicates a broad range of performance among the four gloves tested. The permeation testing also provided several options to the individual making the CPC selection, depending on the amount of protection that needed to be provided over the required time interval. Because enprostil is a drug with known effects on man, broad bioavailability and toxicity data are available. This permits a skin exposure to be made using the permeation data and comparing it to the acceptable dose.

Table 7-12. Permeation data for the mixture propylene carbonate and enprostil against four gloves.

Glove	Observed breakthrough time		Permeation rate	
	Propylene carbonate	Enprostil	Propylene carbonate (µg/min/cm²)	Enprostil (ng/min/cm²)
Playtex	22.7 hr	> 24 hr	4	—
Pharmaseal	46.7 min	300 min	3.7	0.09
Pharmaseal Surgeon's	150 min	450 min	1.3	0.05
Phoenix	32 min	60 min	70.0	72.03

The Actual CPC Selection Process

After completing the preliminary evaluation process, you next must begin the actual selection of the CPC item(s) to use. Where do you find the data? The Introduction to Volume II of this textbook identifies other sources of CPC performance data as well as providing a large database for CPC selection. Up-to-date selection guides from the major manufacturers of CPC should also be kept as reference sources.

But if you are not able to find the CPC performance data for the specific chemical of interest, what is the next step? The major manufacturers and users of the chemical in question should be contacted. The major suppliers of CPC should be contacted for a recommendation or actual permeation testing should be initiated. If these routes still produce no suggestions, Table 7-13 provides general information by class of chemicals for initial evaluation.[28] The table summarizes a number of permeation tests run on six major types of chemical-resistant gloves; polyvinyl alcohol (11–14 mil), latex (4–6 mil), Viton (9–11 mil), nitrile (12–15 mil) butyl rubber (22–27 mil), and neoprene (17–20 mil). This table provides an overview of the polymer's breakthrough time range when the data was organized by family of chemicals tested. It should be used in the CPC selection process as an *initial* guide for experimental planning. When no actual performance data is available to address the CPC selection, the health and safety professional must use his or her professional judgement to make the decision. The safest and recommended way to make CPC selections, especially for moderately toxic to highly toxic materials, is chemical laboratory testing.

Table 7-13. Breakthrough times for six glove types arranged by family of chemicals tested.

Glove material	Breakthrough range	Families tested	Total number tested
Polyvinyl alcohol (11–14 mil)	0–10 minutes	Aliphatic ketone and halocarbon	2
Latex 4–6 mil	" "	Aliphatic amine, alcohol amines, nitriles, and	14
		other functional groups	22
Viton (9–11 mil)	" "	Aldehydes, ethers, epoxides, isocyanates, and	11
		other functional groups	5
Nitrile (1 –15 mil)	" "	Aliphatic halocarbons, and	4
		other functional groups	6
Butyl rubber (22–27 mil)	" "	Aliphatic sulfur, ether, and halocarbon	3
Neoprene (17–20 mil)	" "	Aliphatic isocyanates, hydrocarbon, unsaturated halocarbon	5
Polyvinyl alcohol (11–14 mil)	10–100 minutes	Aliphatic amines, unsaturated aliphatic aldehydes, halocarbons, hydrocarbons and	8
		other functional groups	8
Latex (4–6 mil)	" "	Aliphatic ketones, aliphatic aldehydes	2
Viton (9–11 mil)	" "	Ketones, ethers, and	8
		other functional groups	9
Nitrile (12–15 mil)	" "	Aromatic hydrocarbon, aliphatic amines, aliphatic halocarbons, and	16
		other functional groups	6
Butyl rubber (22–27 mil)	" "	Aromatic hydrocarbons, isocyanates, unsaturated halocarbons, ethers, and	11
		other functional groups	14
Neoprene (17–20 mil)	" "	Aromatic hydrocarbons, nitro, and	9
		other functional groups	37
Polyvinyl alcohol (11–14 mil)	100–200 minutes	None tested	—
Latex (4–6 mil)	" "	None tested	—
Viton (9–11 mil)	" "	Amines, hydrocarbons, epoxides	3
Nitrile (12–15 mil)	" "	Amines, ketones, aliphatic nitro acid salts	5

Table 7-13. (continued.)

Glove material	Breakthrough range	Families tested	Total number tested
Butyl rubber (22–27 mil)	100–200 minutes	Aliphatic amines, halocarbons, and other functional groups	8 6
Neoprene (17–20 mil)	" "	Aliphatic amines, amides, ether, acid ketone	5
Polyvinyl alcohol (11–14 mil)	200–300 minutes	Aliphatic ethers	1
Latex (4–6 mil)	" "	None tested	—
Viton (9–11 mil)	" "	Amines and ether	3
Nitrile (12–15 mil)	" "	Hydrocarbons, halocarbon isocyanates	3
Butyl rubber (22–27 mil)	" "	Aliphatic halocarbons, unsaturated amines	3
Neoprene (17–20 mil)	" "	Aliphatic alcohols, anhydrides	3
Polyvinyl alcohol (11–14 mil)	300–480 minutes	Aliphatic hydrocarbons, ketones, halocarbons, ethers, and other functional groups	22 29
Latex (4–6) mil	" "	Amine salts, salts, alcohol amines, isocyanates, epoxide halocarbons	8
Viton (9–11 mil)	" "	Aliphatic aromatic hydrocarbons and halocarbons, amines, nitriles, halocarbons, alcohols, and other functional groups	48 20
Nitrile (12–15 mil)	" "	Aliphatic amines and halocarbons, hydrocarbons, and other functional groups	13 10
Butyl rubber (22–27 mil)	" "	Aliphatic ketones, aldehydes alcohols, nitriles, amines, acids, and other functional groups	46 28
Neoprene (17–20 mil)	" "	Aliphatic alcohols, amine salts, and other functional groups	5 25

CPC Selection Examples

To familiarize the reader with the basic CPC selection process and the selection information available in Volume II, we will work through a detailed example in this section. The CPC selection worksheet completed in Chapter 9 is also a good reference example. The worksheet provided is meant only as an example that the CPC user can modify to meet program needs. The summary topics on the worksheet are the important items to address. If more space is needed in any of the sections, the sheet format can be expanded to accommodate this. In a well-established CPC program, however, a one-page selection worksheet is possible because of other supporting documents, as well as process description documents that can be referenced for specific information.

Example: 90–92% phenol in water.

Preliminary Workplace Survey

Bulk chemical materials handler.

Process Task Summary

The employee routinely (at least six times a day) transfers hot 122°F (50°C) phenol liquid from a heated 55-gallon drum to an insulated 5-gallon closed bucket, carries it 20 feet, and pours it into a reaction vessel inlet.

Every other day he removes an empty 55-gallon drum and sets up a full 55-gallon drum for this operation. The transfer is done using a forklift. The empty drum is taken to a hazardous chemical waste area where the material handler uses a steam hose to rinse the drum before it is stacked on a pallet for pickup by the chemical supplier. The full drum of phenol is picked up at bulk chemical supply and transferred to the process area where the material handler installs a drain spigot and a bung vent and sets the drum in a horizontal position by using a small overhead crane. A drum heater is placed around the 55-gallon drum to maintain the temperature; the drum is in a ventilated enclosure to maintain the airborne phenol concentration below half the threshold limit value of 5 ppm.

The material handler carries out no additional tasks involving chemicals in the routine eight-hour shift.

Potential or Actual Chemical Hazards

—>90–92% phenol in water.

—Cresols are a typical contaminant in phenols.

—Temperature: room temperature during transfer of drums, 122°F (50°C) during liquid transfer.

—ACGIH TLV = 5 ppm with "Skin " notation; as little as 4.8 g of pure phenol ingested by a person can cause death in ten minutes.

Physical Properties

—Viscous liquid.

—Boiling point 238–288°C.

Potential or Actual Physical Hazards

—Vehicle collisions during forklift operation.

—Thermal burns during the steam cleaning.

—Head and foot hazards for all assignments.

—Lifting hazard when moving full drums around.

—Pinch during drum transfer.

Chemical Contact Periods

—Phenol normally as a liquid or vapor.

—Hand protection; intermittent contact on gloves during transfer and cleanup.

—Feet/boots: continuous, floor routinely contaminated.

—Whole body contact during accident from splash.

Type of Potential Contact

Hand: direct contact of heated liquid during transfer, direct contact of residue during cleanup.

Eye: direct contact from splash during transfer; indirect contact from contaminated items (e.g., gloves).

Feet: direct contact from walking in spilled residue.

Whole body: splash from accidental large spill.

Body Zones of Potential Contact

Eyes, hands, feet: routine potential contact.

Whole body: accidental exposure.

Toxicology of Chemical Exposure[29]

Effects of ingestion: burning sensation in the throat, abdominal pains, increased irritability, headache, absence of corneal reflexes, collapse, convulsions, coma, death.

Effects of skin absorption: local tissue irritation, local tissue necrosis, irregular pulse, darkened urine, stertorous breathing, collapse, vomiting, cold extremities, coma, pallor, cyanosis, convulsions, reduced body temperature, elevated body temperature, dilated pupils, constricted pupils, absence of corneal reflexes, difficulty in swallowing, profuse perspiration, odor of phenol on breath, headache, vertigo, euphoria, dyspnea, ochronosis, general fatigue, local edema, pulmonary edema, abdominal edema, local anaesthesia, damage of kidney tissue, abdominal pain, anemia, depression, liver damage, damage to blood-forming organs, increased irritability, loss of appetite, diarrhea, death.

Effects of inhalation: Similar to effects listed for skin absorption.

Potential Effects of Skin Exposure

—Phenol vapor readily penetrates the skin with an absorption efficiency approximately equal to that for inhalation.[30]

—Liquid phenol in contact with the skin rapidly enters the bloodstream.

—Increases in chemical temperature will increase transport of phenol through the skin.

Permeation, Penetration Degradation Data and Sources

Appendix volume of this textbook.

Recommended Base Materials of Construction for Each or All Components of CPC

Reproduced from Volume II are pages 83 and 84. On the bottom of p. 83 the list of Phenol (carbolic acid) permeation data begins and continues through most of page 84. In addition, immersion weight change test data can be found in Section C, p. 147, summarized here as Table 7-14. There also is one set of data in Section B on permeation data for multi-component chemicals, p. 117.

From the immersion weight change data it can be seen that CPE (chlorinated polyethylene) has a variety of weight changes that vary from 9–68% with chemical immersion over a 24-hr period. Natural rubber and neoprene show observable weight changes over shorter periods of time. All data was collected at 77°F (23°C); in one case the source of the data and thickness are missing. This data indicates that the use of any of these polymers should be based on permeation data, because the phenol contact produces a weight change, which generally means some type of polymer effect.

Table 7-14. Summary of performance tests for phenol, immersion weight change tests.

Chemical name/ CAS No.	Resistant Material	Product Desc. Code	Vendor	Percent Weight Change	Immersion Time (hr)	Temp. (°C)	Thickness (cm)	Ref. No.
Phenol (Carbolic Acid)								
001089520	CPE	060	113	9.10	24.00	23.	.05	204
				68.00	24.00	23.	.05	204
				25.00	24.00	23.	.05	204
	NATURAL RUBBER	001	120	12.00	4.00	23.	.05	236
				2.00	1.00	23.	.05	236
				3.00	.50	23.	.05	236
				2.00	.08	23.	.05	236
	NEOPRENE	010	120	5.00	4.00	23.	.06	236
				1.00	1.00	23.	.06	236
				2.00	.50	23.	.06	236
				2.00	.08	23.	.06	236
Phenol, <30%								
001089521	PE	041	UNK	.20	8,760.00	23.		305
		042	UNK	.10	8,760.00	23.		305
		048	UNK	.20	8,760.00	23.		305

The permeation data on pp. 83–84 in Volume II can be quickly separated by concentration, the >70% section being the most pertinent to the present CPC selection, with the other data useful to note trends to reinforce this data. From the breakthrough times it is clear that butyl, neoprene, and Viton are good candidate materials. From the product descriptions in Section J, starting on page 241, the type of product can be identified and has been noted for this example. For similar evaluations, pages from Volume II can be photocopied and hand-marked, as shown on the following pages.

The obvious easy selections are butyl, neoprene, and Viton gloves. The most current selection guide for the commercial products identified on this chart should be checked to make sure nothing has changed. The glove selection can provide protection for the complete shift or longer. Decontamination options are described in Chapter 8 and should be carefully evaluated or developed for the gloves so that the phenol resistance nature of the glove is not significantly affected.

For body protection, the 6-hr polyethylene (PE) breakthrough time reported for Tyvek makes it a reasonable choice. CPC use directions require the clothing to be changed after each shift or if a significant splash occurs. Various use times for gloves are available, depending on which glove material is picked and how well the decontamination process works. It is also obvious from the data that PVC and nitrile would be a poor barrier material choice. No performance data is available for foot protection, but something other than PVC or nitrile should be selected (e.g., butyl). Standard eye and head protection would be adequate for this application.

The mixture data reported in Section B, page 117, is not pertinent to this particular selection, but readers can examine it to become familiar with this section.

Field Use Considerations

At some point in the selection process, the health and safety professional must integrate all of the performance information to provide the worker with a final list of CPC items. The most difficult part of the process is finding actual permeation data that relates directly to the process in the various manufacturing operations. Berardinelli et al. have carried out an evaluation of the NIOSH CPC portable test method at a Monsanto Chemical Company Plant in Nitro, WV.[31] A wide range of organic materials are used at this plant including acids, bases, alcohols, amines, aliphatics, aromatics, chlorinated hydrocarbons, and halogens. Polyvinylchloride (PVC) flock-lined gloves were used in the plant and surgical latex gloves were used in the lab.

The NIOSH portable system consisted of a personal sampling pump, field permeation cell, and a choice of three detectors: H-Nu Model PI101 photoionization detector; a Foxboro OV 108 total-hydrocarbon flame-ionization detector; or a Photovac Model 10 A10 portable gas chromatograph with a photoionization detector. This system was used at the facility to evaluate a variety of commercial gloves for use in various plant operations. The permeation information was used in conjunction with TLV data to identify mixture components that required specific chemical analysis. Additional gloves made from neoprene and nitrile were added to the CPC program to provide the workers with better chemical protection at specified operations. The results from the portable system should be confirmed with laboratory tests using the ASTM F739 test method. This study illustrates how a comprehensive glove selection program can be carried out at a chemical plant using actual chemical process streams.

Schwope et al. have developed a field evaluation kit to determine chemical permeation by weight loss.[32] A permeation cup was developed that is used to expose the CPC material to the test chemical. A portable balance is used to measure weight loss which can be related to BT and SSPR. A comparison of permeation results for the permeation cup versus the ASTM F739 method is listed in Table 7-15.

Summary of Performance Detail Tests
Permeation Test

Chemical name/ CAS No.	Resistant Material	Product Desc. Code	Vendor	Breakthrough Time (hr)	Permeation Rate (μg/cm²/min)	Temp. (°C)	Thickness (cm)	Ref. No.
001096600	NEOPRENE	018	120	.63	16.03	25.	.05	222
				.33	21.04	25.	.03	222
	NITRILE	005	210	6.00	< .02	23.		080
		019	100	.03	< .02	23.	.04	323
				> 1.00	< 2.00	25.	.04	222
				> 6.00	< .90	23.	.06	107
				> 1.00	< 2.00	25.	.06	222
				> 1.00	< 2.00	25.	.04	222
			503	.09	10.02	25.	.03	222
	NITRILE+PVC	057	210	1.25	90.18	23.		080
		058	100	.18	9.02 - 90.18	23.		107
	PE	006	100	.01	400.80	25.	.01	222
			505	.05	70.14	25.	.01	222
		076	100	.08	90.18 - 901.80	23.		107
	PV ALCOHOL	004	100	> 6.00	< .90	23.		107
		102	100	.25	< .02	23.	.03	323
	PVC	003	120	.01	1,102.20	25.	.01	222
				.01	811.62	25.	.01	222
				.15	100.20	25.	.03	222
				.04	250.50	25.	.02	222
			500	.01	721.44	25.	.01	222
			501	.01	1,603.20	25.	.01	222
				.02	1,603.20	25.	.02	222
		007	210	.33	210.42	23.		080
	SILVER SHIELD	122	118	> 6.00		23.	.01	227
	VITON	009	118	> 8.00		23.	.02	323
				> 8.00		23.	.02	227
Perchloric Acid								
076019030	NATURAL RUBBER	001	210	6.00	< .02	23.		080
	NEOPRENE	002	210	6.00	< .02	23.		080
	NITRILE	005	210	6.00	< .02	23.		080
	NITRILE+PVC	057	210	6.00	< .02	23.		080
		058	100	> 6.00		23.		107
	PE	076	100	> 6.00		23.		107
	PVC	007	210	6.00	< .02	23.		080
		077	100	> 6.00		23.		107
				> 6.00		23.		107
Perchloric Acid, 30-70%								
076019032	NATURAL RUBBER	017	100	> 6.00		23.	.05	107
	NEOPRENE	002	100	> 6.00		23.		107
		018	100	> 6.00		23.	.04	107
	NITRILE	019	100	> 6.00		23.	.06	107
	PVC	007	100	> 6.00		23.		107
Phenol (Carbolic Acid)								
0010895[illegible]	CPE	060	113	3.40		23.	.05	204
				2.92	60.12	23.	.05	204
	NATURAL RUBBER	001	210	.58		23.		080
		017	100	> 1.00	< 3.01	25.	.03	222
				1.00	9.02 - 90.18	23.	.05	107
			120	.27	[illegible]5.03	25.	.02	222

Summary of Performance Detail Tests
Permeation Test

Chemical name/ CAS No.	Resistant Material	Product Desc. Code	Vendor		Breakthrough Time (hr)			Permeation Rate (µg/cm²/min)	Temp. (°C)	Thickness (cm)	Ref. No.
001089520	NATURAL RUBBER	017	502	>	1.67		<	3.01	25.	.05	222
			504	>	1.00		<	3.01	25.	.05	222
				>	1.00		<	3.01	25.	.06	222
	NEOP+NAT RUBBER	026	102	>	1.00		<	3.01	25.	.05	222
	NEOP/NAT RUBBER	008	114	>	1.00		<	3.01	25.	.05	222
	NEOPRENE	002	100	>	6.50		<	.90	23.		107
				>	1.65		<	3.01	25.	.08	222
			210		.67				23.		080
		018	100		3.00	9.02	-	90.18	23.	.04	107
			118	>	1.00		<	3.01	25.	.08	222
			120	>	1.00		<	3.01	25.	.05	222
				>	1.00		<	3.01	25.	.07	222
				>	1.00		<	3.01	25.	.05	222
				>	1.00		<	3.01	25.	.03	222
	NITRILE	005	210		.67				23.		080
		019	100		.93			300.60	25.	.04	222
				>	1.00		<	3.01	25.	.06	222
					.53			300.60	25.	.04	222
			503		.60		>	250.50	25.	.03	222
	NITRILE+PVC	057	210		2.00				23.		080
	PE	006	100	>	1.00		<	3.01	25.	.01	222
			505		1.00			3.01	25.	.01	222
	PV ALCOHOL	004	100		.50	9.02	-	90.18	23.		107
	PVC	003	120		.05			190.38	25.	.01	222
					.13			120.24	25.	.01	222
					.53			77.15	25.	.03	222
					.25			100.20	25.	.02	222
			500		.10			130.26	25.	.01	222
			501		.10			120.24	25.	.01	222
					.06			120.24	25.	.02	222
		007	100		1.25	.90	-	9.02	23.		107
			210		1.33				23.		080
	TEFLON	069	510	>	3.00		<	.02	23.	.05	303
Phenol, >70% 001089523	BUTYL	014	118	>	20.00				23.	.06	323
				>	20.00				23.	.04	227
	NEOPRENE	018	100	>	10.67				23.	.05	000
		125	103				<	.02	23.		045
	NITRILE	019	103					18.04	23.		045
			118		.58			1,274.54	23.	.03	323
					.65		>	9,018.00	23.	.04	227
	NITRILE+PVC	058	100		.83	.90	-	9.02	23.		107
	PE	076	100		6.00		<	.90	23.		107
	PVC	007	103					18.04	23.		045
		077	100		.50	.90	-	9.02	23.		107
					1.50	.90	-	9.02	23.		107
	VITON	009	118	>	15.00				23.	.03	323
				>	15.00		<	.02	23.	.02	227
Phenolphthalein 000770980	NATURAL RUBBER	017	506	>	8.00				23.	.02	323
	NEOPRENE	018	100	>	8.00				23.	.04	323

Table 7-15. Test results (means) for butyl-coated nylon.

	Permeation Cup		ASTM F739-85	
	BT[a]	PR[b]	BT[a]	PR[b]
Acetone	ND[c]	ND	ND	ND
Hexane	30	55	13	83
Methanol	ND	ND	ND	ND
Toluene	7	290	7	528
Acetone/hexane (25/75, v/v)	7	250	4	465
Acetone/hexane (50/50, v/v)	6	220	6	330
Acetone/hexane (75/25, v/v)	13	80	16	92

[a] Breakthrough time, min.
[b] Steady-state permeation rate, μg/cm² min.
[c] No breakthrough detected in 1 hr for the cup test or 6 hr for ASTM F739-8.

These results formed the basis for the experimenters to identify gravimetric methods as an expedient means for providing field personnel with considerably more information than is currently available to them on the chemical resistance of CPC.

Gravimetric permeation field test kits are now commercially available from Texas Research Institute, Inc. and Arthur D. Little, Inc. Figure 7-9 shows the contents of the TRI kit.[33] The kit components are:

- Field rugged, portable analytical balance (accurate to ± 0.01g).
- Three permeation cups.
- Three permeation cup stands.
- A template kit with stencil, scissors, and marker.
- A Teflon beaker.
- A graduated cylinder (10 ml).
- A calibrated weight (10g).
- A set of instructions and data forms.
- A carrying case.

To conduct a test, the health and safety professional obtains a test specimen and weighs it, then places the test chemical in the test cup with the lid clamped around the specimen; the test cup is weighed and noted as weight at time 0. Additional weighings are made at a series of times such as 3, 6, 9, 12, 15, 20, 25, 30, 40, 50, and 60 minutes. The data analysis is simple, where three successive weight losses occur, the first time at which the loss is observed is called the BT for the material/chemical pair. If the permeation rate is approximately constant during the last weight measurements (within 10–20%) it is said to be a steady-state permeation rate. If the rate continues to increase without any leveling off, it is called a maximum observed permeation rate. The cell is disassembled and the test specimen blotted dry and weighed. By comparing the weight before and after exposure, any increase can be taken as an indication of degradation.

The ability to carry out field test evaluations on CPC with the actual chemical(s) that will be encountered is very valuable. The results represent a good measure of actual CPC performance under realistic field conditions. The method is limited in its accuracy due to the analytical balance sensitivity and field conditions. The method does take time and requires a trained technician to carry it out accurately. By weighing the pros and cons of the gravimetric permeation field test method, it is obvious that when properly used it is a very valuable tool for fast and accurate selection of CPC in the field.

Another aid to quick selection of CPC in the field is the *Quick Selection Guide to Chemical Protective Clothing* by Forsberg and Mansdorf.[34] This pocket-sized booklet provides an easy-to-use color-coded selection system for 420 hazardous chemicals against 11 generic CPC materials. With a short training class this guide could also be provided to workers to aid them in their own CPC selection process.

Another useful procedure to keep in mind when implementing a CPC program in the field is double gloving or even double suiting. With two

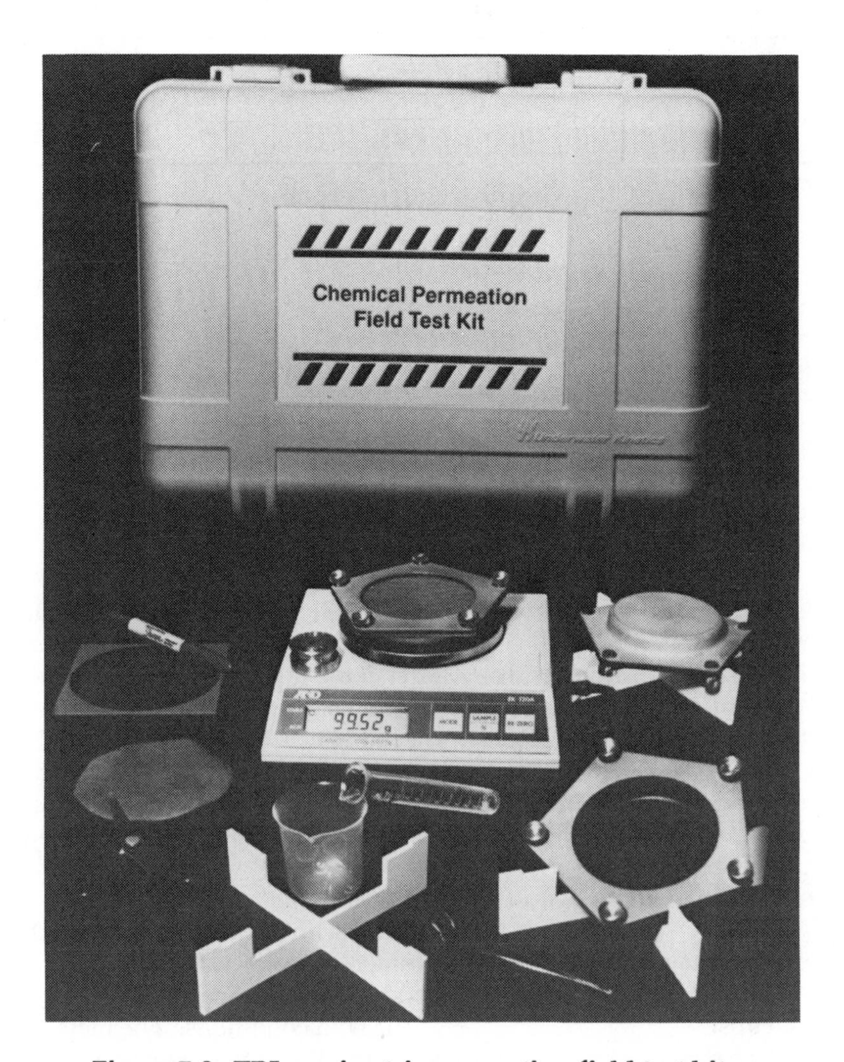

Figure 7-9. TRI gravimetric permeation field test kit.

gloves, the best properties of one glove can be used, while the worst properties can be addressed by the second glove. If your program uses many different gloves, consider color-coding the available gloves by sole-sourcing specific brands. This simple approach to glove control provides easy selection by the employees and easy auditing by the health and safety staff.

Chapter 8 provides detailed treatment of the decontamination of CPC. This topic should be an intimate part of any CPC program, with decontamination procedures discussed and selected as part of the initial evaluation process. Other field use considerations—such as durability, flexibility, duration of use, compatibility with other equipment, and disposal requirements—must also be examined as part of the original selection process.

If CPC is provided for emergency response or as part of a hazardous waste cleanup site, Federal legislation 29CFR Part 1910.120 requires a personal protective equipment program. Requirements similar to those of the respirator standards are specified. These types of requirements can be expected to expand and grow, requiring a higher level of attention being paid to CPC programs and use of protective materials.

Summary

The selection of CPC is an assignment that must be carried out carefully and diligently. A selection worksheet has been organized and outlined to provide consistency in the CPC selection process. By completing all of the topics noted, a good database will be available to permit the health and safety professional to make an informed CPC selection. Because the data analysis in many cases does not match the exact conditions of the workplace, professional judgement must be relied on. Follow-up monitoring of the CPC user will assure that the selection was correct.

Such variabilities as temperature, pressure, polymer thickness, chemical mixtures, variability of polymer composition, and chemical contact time have been shown to affect CPC performance. The CPC selection process must, therefore, take these variabilities into consideration.

One of the most significant tasks, but most difficult at present, is the determination of the toxicity of the chemical and specifically whether it will permeate through the skin. For the most part, the determination of skin toxicity is made on a comparative basis, which provides the opportunity for ultra conservatism or blatant disregard. The selection process is one of making many decisions on general performance data, or on performance data at the wrong temperature, etc. The need to be ready to carry out one's own permeation tests to support the decision process is a reality for a CPC program of any size. By using the information presented in this chapter and the appendix volume of this textbook, the health and safety professional has been provided adequate information and an understanding of the selection process to be ready to make his or her own choices.

References

1. American Conference of Governmental Industrial Hygienists, *Threshold Limit Values and Biological Exposure Indices for 1988-1989*, American Conference of Governmental Industrial Hygienists, Cincinnati, OH (1989).

2. P. Grandjean, "Preventing Percutaneous Absorption of Industrial Chemicals: The 'Skin' Notation," *Am. J. Ind. Med.*, 14, 97 (1988).

3. S.S. Mileti, *Chemical/Physical Properties Guidelines for Du Pont*, "Skin Notation," American Industrial Hygiene Conference, Montreal, Canada, Abstract #171 (1987).

4. U.K. Brown, *Acute Toxicity in Theory and Practice,* John Wiley and Sons, Inc. , New York (1986), p. 3.

5. National Institute for Occupational Safety and Health, *The Industrial Environment—Its Evaluation and Control*, US Government Printing Office, Washington, DC, p. 63 (1973).

6. G. Salvend, *Handbook of Human Factors*, Chapter 6.6, "Ergonomic Factors in Chemical Hazard Control," John Wiley and Sons, New York (1987).

7. J. Doull, C.D. Klaassen, and M.O. Amdur, *Casarett and Doull's Toxicology*, 2nd ed., Macmillan Publishing Co., New York, p. 12 (1980).

8. D.N. Eiser, AT&T Microelectronics, Reading, PA, private communication.

9. J.F. Stampfer, M.J. McLeod, M.R. Betts, A.M. Martinez, and S. Berardinelli, "Permeation of Eleven Protective Garment Materials by Four Organic Solvents," *Am. Ind. Hyg. Assoc. J.* **45**, 642 (1984).

10. I. Drabaek and C. Ursin, "Feasibility of Using Radioactive Tracers for Studies of Permeation of Chemicals Through Protective Clothing Materials, Final Report," the Danish Isotope Center, document 438.70 (1987).

11. C. Ursin and I. Drabaek, "Carbon-14 Tracers in Permeation Studies, Feasibility Demonstration," in *Performance of Protective Clothing: Second Symposium*, ASTM STP 989, S. Z. Mansdorf, R. Sager and A. R. Nielson, Eds., American Society for Testing and Materials, Philadelphia, PA (1988).

12. J. Harville and S.S. QueHee, "Permeation of a 2,4-10 Isooctyl Ester Formulation Through Neoprene, Nitrile and Tyvek Protection Materials," *Am. Ind. Hyg. Assoc. J.* **56**, 438 (1989).

13. G.O. Nelson, B. Lum, G. Graham, C. Wong and J. Johnson, "Glove Permeation by Organic Solvents," *Am. Ind. Hyg. Assoc. J.* **42**, 217 (1981).

14. R.W. Weeks and B.J. Dean, "Permeation of Methanolic Aromatic Amine Solutions Through Commercially Available Glove Materials," *Am. Ind. Hyg. Assoc. J.* **39**, 169 (1971).

15. J.O. Stull, *Development of a U.S. Coast Guard Chemical Response Suit*, National Technical Information Service, CG-D-16-87, p. 19 (1987).

16. K. Forsberg, A. Lennarson, K. Olssen, "Development of Safety Gloves for Printers," Royal Institute of Technology, Stockholm, Sweden [Translated by National Institute for Occupational Safety and Health, Cincinnati, OH], (1981).

17. J.L. Perkins, "Chemical Protective Clothing: I. Selection and Use Considerations," *Appl. Ind. Hyg.* **2**, 222 (1987).

18. E.C. Gunderson, B.A. Kingsley, C.L. Wilham, and D.C. Bomberger, "A Practical Study in Laboratory and Workplace Testing," presented at American Industrial Hygiene Conference, Montreal, Canada, Abstract #268 (1987).

19. K. Forsberg, AGA, AB, Lidingo, Sweden, private communication.

20. J. Eggestad and B.A. Johnson, "Penetration of Permeable Chemical Protective Clothing by Drops Under Pressure Loads; Repellent and Wicking Fabrics," *Am. Ind. Hyg. Assoc. J.* **50**, 391 (1989).

21. C.N. Schlatter and D.J. Muller, "The Influence of Film Thickness on Permeation Resistance Properties and Unsupported Glove Films," in *Performance of Protective Clothing*, ASTM STP 900, R.B. Baker and G.C. Coletta Eds., American Society for Testing and Materials, Philadelphia, PA (1986).

22. K. Forsberg and S. Faniadra, "The Permeation of Multi-Component Liquids Through New and Pre-Exposed Glove Material," *Am. Ind. Hyg. Assoc. J.* **47**, 189 (1986).

23. R.L. Mickelson, M.M. Roder and S.P. Berardinelli, "Permeation of Chemical Protective Clothing by Three Binary Solvent Mixtures," *Am. Ind. Hyg. Assoc. J.* **47**, 236 (1986).

24. R.L. Mickelson and R.C. Hall, "A Breakthrough Time Comparison of Nitrile and Neoprene Glove Materials Produced by Different Manufacturers," *Am. Ind. Hyg. Assoc. J.* **48**, 941 (1987).

25. V.I. Man, V. Bastecki, G. Vondal and A.P. Bentz, "Permeation of Protective Clothing Materials: Comparison of Liquid Contact, Liquid Splashes and Vapors on Breakthrough Times," *Am. Ind. Hyg. Assoc. J.* **48**, 551 (1987).

26. S.P. Berardinelli and E.S. Moyer, "Methyl Isocyanate Liquid and Vapor Permeation Through Selected Respirator Diaphragms and Chemical Protective Clothing," *Am. Ind. Hyg. Assoc. J.* **48**, 324 (1987).

27. M. Conoley and D.B. Walters, "Cumulative Permeation Test Results for the National Toxicology Program," presented at American Industrial Hygiene Conference, Dallas, TX, Abstract #96 (1986).

28. T. Bryoning and T.J. Massey, "Permeation of Enprostil and Propylene Carbonate Through Commercially Available Glove Material, " presented at the American Industrial Hygiene Conference, San Francisco, CA (1989).

29. National Institute for Occupational Safety and Health, "Criteria for a Recommended Standard...Occupational Exposure to Phenol," US Government Printing Office NIOSH 76-196, Washington, DC (1976).

30. S.K. Protrowski, "Evaluation of Exposure to Phenol—Absorption of Phenol Vapor in the Lungs and through the Skin and Excretion of Phenol in Urine," *Br. J. Ind. Med.* **28**, 172 (1971).

31. S.P. Berardinelli, R.A. Rusczek, and R.L. Mickelsen, "A Portable Chemical Protective Clothing Test Method: Application at a Chemical Plant," *Am. Ind. Hyg. Assoc. J.* **48**, 804 (1987).

32. A.D. Schwope, T.R. Carroll, R. Huang, and M.D. Royer, "Test Kit for Field Evaluation of the Chemical Resistance of Protective Clothing," *Performance of Protective Clothing and Materials: Second Symposium*, ASTM STP 989, S.Z. Mansdorf, R. Sager, and A.P. Nielsen, Eds., American Society for Testing and Materials, Philadelphia, PA (1988).

33. J.O. Stull, "A New Permeation Test Kit to Aid Field Selection of Chemical Protective Clothing," Information Article, Texas Research Institute, Austin, TX (1989).

34. K. Forsberg and S.Z. Mansdorf, *Quick Selection Guide to Chemical Protective Clothing,* Van Nostrand Reinhold, New York (1989).

Chapter 8

Decontamination of Protective Clothing

C. Nelson Schlatter

Introduction: When Should Garments Be Decontaminated?

When an article of protective clothing becomes contaminated with chemicals, it should either be discarded or decontaminated. The better choice in any given application depends on a variety of factors. Discarding contaminated clothing may be the easiest, but most expensive, option. Decontamination can be an attractive option if it helps to achieve optimum protection for workers at the lowest possible cost.

The range of replacement costs for chemical protective clothing is broad. The current price list of Edmont, a major glove manufacturer, lists gloves from 2.5¢ to $17.41 per pair. Inexpensive items from all manufacturers may be obviously marketed as "disposables," but the price range for chemical protective clothing has no gap that clearly separates "disposable" from "reusable" items. Even garments marketed as "reusable" are frequently thrown out after a single use.

The relative costs of replacement and decontamination depend on the individual situation. In some instances, only an expensive garment will provide adequate protection against the chemicals in the workplace. The high replacement cost of such garments will increase the financial benefit of developing a successful recycling and decontamination process.

The cost of the actual cleaning process may not be the most important factor in decontamination. For example, any attempt to clean gloves used in polychlorinated biphenyl (PCB) will generate PCB-contaminated cleaning solution. This waste solution will probably be harder to handle and cost more for proper disposal than the contaminated gloves, since the gloves are solid and probably much less bulky than the cleaning solution. Replacing PCB-contaminated gloves after each use is generally the least expensive alternative.

In addition to relative cost of decontamination, the toxicity of any chemical residue still remaining on the garment after decontamination can be important. It can be very difficult to remove the last traces of contaminant, and if trace quantities may cause major health problems, disposal of the garment is probably the better approach. On the other hand, some chemical protective clothing is used to protect only against such mild chemicals as sodium chloride brine (salt water), and a trace residue of brine is totally insignificant from a health standpoint.

The previous examples can also illustrate how the ease of cleaning can affect decontamination. The PCB liquid is a very sticky and persistent oil that is not completely removed even after multiple cleanings in water, detergent solution, and isooctane, followed by baking in an oven overnight.[1] Sodium chloride brine, on the other hand, is readily water soluble and can easily be washed out of protective garments.

Finally, the endurance of protective garments is an important factor. Rubber and plastic materials may fail subtly by losing permeation resistance, or more obviously by developing cracks or pores, by becoming stiff, or by suffering other deleterious changes in physical properties. These changes may be caused by exposure to the hazardous chemical, by exposure to solvents or other materials used in the decontamination process, or by heat or other environmental factors that may be encountered in storage, use, or decontamination. Recycled garments should be inspected for cracks, swelling, color change, or other evidence of possible problems before being reused.

Chemical Contaminants—Properties That Affect Decontamination

The effectiveness of decontamination is influenced both by the nature of the polymers from which the garment is made and by the properties of the contaminating chemical itself. Since polymer properties are discussed in some detail in Chapter 3, this chapter will concentrate instead on the contaminating chemicals. The properties that most strongly affect decontamination are physical state and solubility parameters.

Physical state determines how intimately the contaminant comes into contact with the protective garment, and thus has a direct influence on the ease of decontamination. From this point of view, sticky viscous liquids and buttery solids are the worst sort of contaminants, followed by free-flowing liquids, then solid powders, then gases, and finally hard solids (which can generally be handled without contaminating the protective garment at all).

Solubility parameters are convenient tools for predicting which chemicals are compatible and will dissolve in each other. These parameters are a mathematical formalization of a generalization that chemists have known for decades, that "like dissolves like."

Ideally, a contaminating chemical will have solubility parameters very different from those of the protective polymers and very similar to those of the decontaminating solvent. If these conditions are met, the contaminant will have a minimal effect on the protective garment, and the decontamination process can be very straightforward.

It may still be possible to use and decontaminate a garment even if these conditions are not fully met; in this case, however, the garment may need frequent changing and cleaning to reduce the amount of chemical that permeates through to the wearer. A more vigorous decontamination process may be needed as well.

The concept of the solubility parameter was originally developed by Hildebrand.[2] Hansen later extended this work and resolved the Hildebrand parameter for any organic chemical into three independent components related to its polarizability, dipole moment, and hydrogen bonding.[3,4] Most common organic chemicals have polarizability parameters in the relatively limited range of 14 ± 2 $MPa^{-1/2}$. Differences in this narrow range have little effect on compatibility. It is therefore possible to plot the other two components—the dipole parameter and the hydrogen-bond parameter—against each other on ordinary graph paper and get a good indication of which chemicals are compatible. If the points for two solvents on such a graph are close, the solvents will probably dissolve in each other.

If the solubility parameters of common CPC materials are added to such a plot, it should also be possible to predict which chemicals will permeate through which materials. A researcher at the Danish Institute for Industrial Environment has written a paper in which this was done.[5] Additional work has shown that the placement of the equal-chemical-resistance curves reported in this paper may need to be refined, but it is a worthwhile start on a simple system for estimating chemical resistance before running actual tests.[6]

The *CRC Handbook of Solubility Parameters and Other Cohesion Parameters* gives the derivation and uses of solubility parameters in much greater detail.[7]

Permeation and Decontamination: The Trapping Effect

A molecule that permeates into a protective garment can end up in several places. It can permeate all the way through and reach the inside surface, it can remain dissolved within the polymer film, or it can be trapped in a reservoir that may exist within the garment. In the second and third cases, a decontamination process that cleans the surface of the garment may not remove the trapped chemical. Obendorf and Solbrig have reported that organophosphate pesticides remain trapped in the hollow lumens of cotton fabric even after washing in Tide, the most widely used home laundry detergent.[8] In some of my own research, I found a residue of xylene in nitrile and a residue of hexane in neoprene after exposure to the respective chemicals and hot caustic washing.[9] The pesticides tested by Obendorf and Solbrig were tagged with osmium tetroxide and identified in place by energy dispersive x-ray analysis, so they may have been permanently trapped and therefore unable to affect anyone in contact with the fabric; however, I detected the xylene and hexane during permeation tests after washing. The trapped contaminant caused an early breakthrough followed by a low plateau in the permeation rate until the normal breakthrough time. The permeation rate then began to climb, as it does when brand new gloves are tested.

Figure 8-1 is a comparison plot of selected test results for xylene versus nitrile. The early breakthrough and low initial plateau of the exposed and decontaminated sample can be seen. A second change is the lower maximum permeation rate after decontamination, a change that was reproducible for other glove-solvent combinations but which is not yet well understood; it may be due to changes in the morphology of the glove materials during exposure and decontamination.

If a worker had been relying on recycled glove materials for complete protection against xylene or hexane, he or she would have been exposed much sooner than would be predicted based on laboratory testing of brand new gloves, but he or she might have a lower total dose for longer-term exposures. The same effects may be observed for other more toxic chemicals as well.

In order to achieve complete protection, any decontamination process must remove or destroy both surface contamination and also any hazardous material that has entered the garment matrix. For many chemicals, this is not an easy goal.

Decontamination by Aeration

The simplest way to decontaminate a garment is to place it in a stream of warm air for several hours and allow the contaminant to evaporate. Obviously, this works best when the contaminant is highly volatile. Rubber medical equipment, such as tubes and catheters, is routinely sterilized by highly volatile ethylene oxide and, after decontamination by aeration, is then

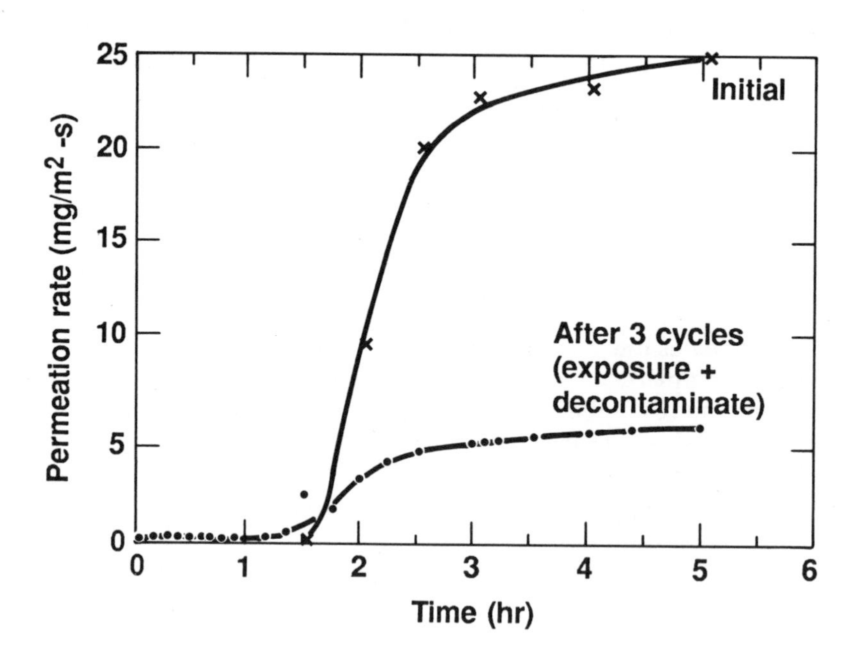

Figure 8-1. Comparison plot of selected test results for xylene vs nitrile.

used in direct contact with the bloodstream without materially harming the patient or violating FDA regulations.

To sterilize rubber equipment with ethylene oxide, a medical technician typically treats it first for 20 to 60 minutes with steam under a vacuum of about 28 inches of mercury (about 7 kPa residual pressure). This improves gas absorption in the following step, exposure to about 0.1–0.2 atm (10–20 kPa) of ethylene oxide pressure for two to six hours. A partial vacuum may be maintained in the chamber, or the ethylene oxide may be diluted by another gas, such as Freon.

Decontamination by aeration follows. Typically, the chamber pressure is cycled several times from atmospheric to as low as 0.1 atm. The medical equipment is then placed in a room where the air is changed about 20 to 30 times per hour and is left until the residual ethylene oxide is below the FDA tolerance limit of 1 ppm. If the aeration room is heated to 130°F (55°C), this process can be completed in 8 hr; if the room is maintained at 80°F (27°C), the process may take 2–3 days.

The gloves in my own experiments mentioned above went through a 150°F caustic wash and subsequent drying in a hot air oven for another hour at 150°F. The test gloves still retained some xylene and hexane. Aeration times as long as 16 hr were also tried, but only at room temperature in the air stream of a ventilation hood. Again, I found indications that solvents were retained in the glove rubber.

Recent work at the Lawrence Livermore National Laboratory has shown that eight common industrial solvents from various chemical classes can be completely removed from a commercial butyl suit fabric by heating for eight hours at 50°C in a lab oven.[10] Eight hours in a hood at 25°C left solvent residues of up to 6%. This confirms that aeration can be an effective decontamination method if it is run for a long enough time at a high enough temperature. It demonstrates also that aeration works for isopropyl alcohol, acetonitrile, chlorobenzene, and several other chemicals that are significantly less volatile than ethylene oxide.

A recent study at Du Pont included aeration decontamination trials with nitrobenzene and several protective garment materials.[11] A two-second splash induced measurable contamination by permeation into several of the garment materials tested, especially neoprene. The splashed-in nitrobenzene was totally removed from the neoprene by aerating for 24 hours at 70°C.

Many questions still remain about decontamination by aeration. How is the minimum effective aeration time altered by changes in conditions during contamination, by aeration temperature, by garment thickness, and by choice of garment polymer? How volatile must the contaminant be for the process to be successful? Will repeated aeration harm the properties of protective garments? Is it necessary to remove the trace chemicals from the heated air used in the aeration process? We may expect to see additional papers on aeration in the scientific literature over the next few years.

Water-Based Decontamination

The most commonly used decontaminating solvent is water. Water is by far the most readily available of all possible solvents, it generates no toxic fumes or contamination of its own, it has a minimal effect on the physical properties of most protective clothing materials, and machines for automatic water washing and rinsing can be found in most houses in the United States.

The Edmont recommendations for cleaning most of their manufactured supported gloves are as follows:

- Use commercial laundry soap or detergent.
- Use water at 180°F.
- Wash for 10 minutes.
- Rinse in 180°F water.
- Repeat wash and rinse, if necessary.
- Rinse in cold water.
- Tumble dry, at a maximum temperature of 160°F.

This process was originally intended to remove grease, dirt, and soil, but should remove most chemical contamination as well. According to tests with pesticides, residues can be minimized by using very hot water, with few garments and the maximum water level in each washing machine load.[12]

Another approach to water-based decontamination is used at spill sites by Environmental Protection Agency cleanup teams. Workers at these sites frequently wear multiple layers of protective clothing. When leaving the contaminated area, they must scrub each layer with a long-handled, soft-bristled scrub brush and copious amounts of detergent/water or other decontaminating solution, then rinse with copious amounts of water; they then remove one layer of clothing and begin again on the next. The last step is a shower to clean the worker's skin. To prevent cross-contamination, wash water should not be used more than once in this process. The only equipment needed for this approach is scrub brushes, drums to carry the used garments away, tanks of detergent/water or decontaminating solution, and children's wading pools or something similar to catch the contaminated wash and rinse water. Such simplicity can be a real advantage at a remote spill site.

Once again, complete decontamination cannot be guaranteed. However, the goals of keeping the worker clean as he or she removes the protective clothing, as well as keeping chemicals from being carried out of the polluted area, are at least as important as the goal of recycling the protective garments themselves.

Another example of the "scrub and rinse" approach is the plan for decontamination in the event of a chemical warfare attack. The US Army plans to have soldiers scrub their chemical warfare uniforms and equipment with a strong chlorine bleach solution to decompose the toxic agents. This will presumably be effective against whatever chemical agents an enemy may be using.

Decontamination Based on Organic Solvents

Any organic solvent can be used as a decontaminating solution if the contaminant is soluble in it. However, an ideal general-purpose decontaminating solvent would also have certain specific properties: (1) no permanent effects on the physical properties of the garment; (2) low cost; (3) stability through multiple cleaning and redistillation cycles; (4) low heat of vaporization (to save energy when solvent is removed from the garments and when the used solvent is redistilled); and (5) low flammability and toxicity. The ideal solvent does not exist, but Stoddard solvent, perchloroethylene, and Freon TF are all close enough to justify the development of dry-cleaning processes for decontamination. Their properties are compared in Table 8-1.

Stoddard solvent and perchloroethylene are by far the most widely used solvents in the general dry cleaning industry; Freon TF is generally reserved for specialty applications. However, the dry cleaning of chemical protective clothing is definitely a specialized process, and the only people now doing this use Freon TF.

One commercial application of dry-cleaning protective clothing involves natural rubber gloves used in the nuclear power industry. During use, gloves are contaminated with low levels of particulate radioactivity; they must be cleaned to avoid very expensive disposal in a radioactive waste dump. This decontamination is done commercially using Freon TF in a Vic model 478 dry-cleaning machine.[13]

As an experiment, I had Edmont natural rubber and nitrile gloves dry cleaned by this process, and then checked their permeation and physical properties.[14] They were unaffected. In contrast, a regular commercial dry cleaning using perchloroethylene ruined the natural rubber gloves (Figure 8-2) and left an appreciable

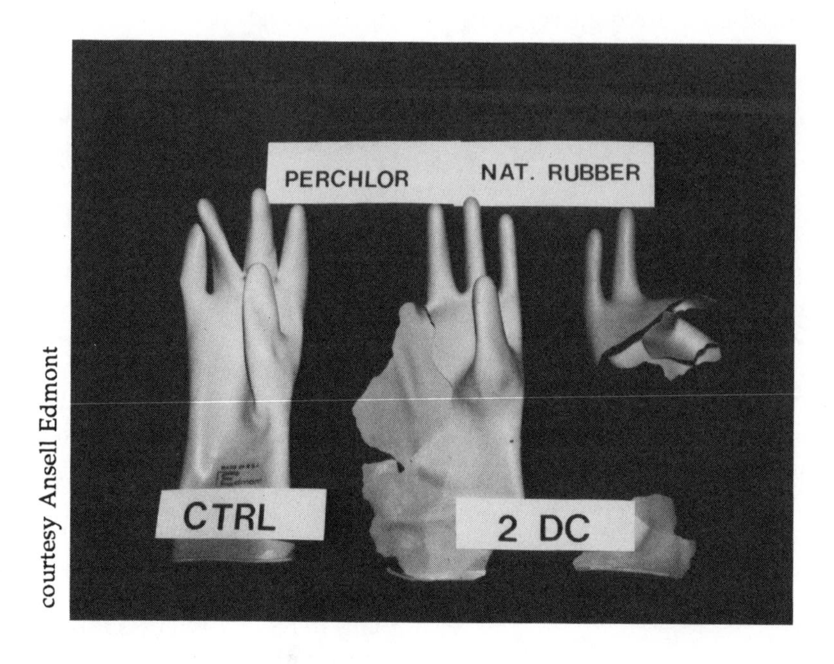

courtesy Ansell Edmont

Figure 8-2. Damage done to natural rubber gloves from a commercial dry-cleaning process using perchloroethylene.

residue of perchloroethylene in the nitrile gloves. The use of a specialty solvent seems to be justified in this application.

These results would not have been predicted based on laboratory degradation testing. According to Table 8-1, natural rubber degrades severely and is "Not Recommended" against either perchloroethylene or Freon TF. Apparently the gloves recover their physical properties at the end of the process when the solvent is removed. The Freon-based process handles the gloves gently during cleaning so that they are intact when the full physical properties are

Table 8-1. Properties of common dry cleaning solvents.

	Stoddard Solvent	Perchloroethylene	Freon TF
Cost	Very low	Low	Moderate
Multi-cycle use	No	Yes	Yes
Heat of vaporization	—	Low	Very low
Solvent power	Depends on the contaminant present in the protective clothing.		
Flammability	Moderate	None	None
Now in common use?	Yes	Yes	No
Toxicity: 16			
LD_{50}	—	2100 mg/kg	43 g/kg
LD_{lo}	1470 mg/kg	2200 mg/kg[a]	—
Degradation 17[b] Rating, vs:			
Nitrile	E	G	E
Neoprene	E	NR	G
PVC	E	NR	NR
PVA	E	E	G
Nat. rubber	NR	NR	NR

[a] This value is for subcutaneous injection. A value for direct injection to the bloodstream is not reported here due to lack of realism. With this exception, these are the smallest LD_{50} and LD_{lo} values from the 6th edition of Sax.

[b] Edmont tests for degradation by immersing samples of protective polymer films in solvents or other chemicals for 30 minutes, then observing any changes in sample size, weight, or appearance. The protective polymers are then rated Excellent, Good, Fair, Poor, or Not Recommended against each chemical. For a dry cleaning solvent, a low rating implies a high risk of damaging protective garments during dry cleaning.

restored in drying. Perchloroethylene-cleaned natural rubber gloves also recover a large fraction of their physical properties, but they are torn so badly during cleaning that this recovery is of academic interest only.

Recovery of physical properties is not always observed, though. CPC materials may be divided into two classes, rubber and plastics. Unlike rubber materials, which are inherently flexible, plastics must be made flexible by the addition of plasticizers (see Chapter 3). Many common plasticizers are low-molecular-weight compounds that may dissolve in contaminating chemicals or in dry-cleaning solvents. The garments shrink and stiffen when this occurs, and there is no practical way to restore them for reuse.

A second experiment involving Freon was run on firefighter's turnout gear after it had been used to fight a major PCB fire.[18] Freon dry cleaning removed 95% of the PCB from the garments, and 99.9% of the PCB from the same garments soiled experimentally in the laboratory. In contrast, water-based cleaning using detergent, buckets, and scrub brushes merely redistributed the PCB more evenly over the garments due to contamination of the soapy water.

The recent study at Du Pont included decontamination trials of nitrobenzene with Freon.[11] Two ten-minute washes did not remove all of the nitrobenzene from butyl, neoprene, or Viton garment material, and breakthrough times for nitrobenzene after Freon decontamination were much shorter. In my own work, I tested nitrile and neoprene versus acetone, isobutyl alcohol, and Freon itself, and I found no significant changes in subsequent breakthrough times after dry cleaning with Freon.[14] The differing results may be specific to the combinations of challenge chemical and protective film that were tested.

Summary and Conclusions

Which chemicals can be adequately removed from which protective garment materials, and how? Currently we have no complete answers.

Part of the problem is that we have not defined what we mean by "adequately." How much glycol ether, for example, can be retained in a recycled natural rubber glove after decontamination without posing an excessive risk to the worker? The answer to that clearly depends on the safe skin-contact levels and on the rate of release of trapped glycol ether from rubber. Neither of these parameters has yet been determined. Obviously, more work must be done.

In addition, we have no good method to measure residues in protective clothing to learn if an adequate reduction has been achieved. Such a method should be non-destructive and should clearly identify and quantify the residue. This poses a difficult problem for analytical chemistry. The best methods available to date depend on cutting samples from the protective film, and although this may be adequate for some research projects, it ends all hope of recycling for the garment.

To summarize the decontamination processes:

- The simplest process is aeration, which can be effective enough even for highly critical medical applications if the specific contaminant is sufficiently volatile.

- Water/detergent washing is generally effective in removing a high percentage of surface contamination. Chemical residues may not be completely removed, however, especially if they are contained within the clothing matrix and not on its surface. Cross-contamination can be a problem if wash water is used repeatedly, which can easily happen in manual bucket-and-brush processes.

• Water/bleach washing is considered by the US Army for decontamination of chemical warfare agents. This method should be considered if the contaminant can be oxidized.

• Freon-based dry cleaning has also proven to be effective. Unlike water-based processes that can use buckets and brushes or ordinary washing machines, specialized equipment is needed for dry cleaning. This method may be more effective though, since the specialized equipment is designed to minimize cross-contamination.

References

1. J.F. Stampfer, M.J. McLeod, M.R. Betts, A.M. Martinez, and S.P. Berardinelli, "Permeation of Polychlorinated Biphenyls and Solutions of These Substances Through Selected Protective Clothing Materials," *Am. Ind. Hyg. Assoc. J.* **45**: 634 (1984).

2. J.H. Hildebrand, J.M. Prausnitz, and R.L. Scott, *Regular and Related Solutions*, (Van Nostrand Reinhold, Princeton, NJ, 1970).

3. C.M. Hansen, "The Universality of the Solubility Parameter," *Ind. Eng. Chem. Prod. Res. Dev.* **8**: 2 (1969).

4. C.M. Hansen, "Solvents for Coatings," *Chem. Tech.* **2**, 547 (1972).

5. U.L. Christensen, *Handsker-sikre/usikre; En hjaelp til en vurdering af gennemtraengelighed, Institut for Arbejdsmiljø*, Denmarks tekniske Højskole (1983) (English translation: "Gloves—Safe/Unsafe; An Aid for the Evaluation of Permeability," TR 84-0194, SCITRAN, Santa Barbara, CA).

6. C.N. Schlatter, "Glove and Solubility Parameters; Supplemental Experiments for Review of a Technical Article," presented at the meeting of ASTM committee on F23 on Protective Clothing, Washington, DC, June 1985.

7. A.F.M. Barton, *CRC Handbook of Solubility Parameters and Other Cohesion Parameters* (CRC Press, Boca Raton, FL, 1983).

8. S.K. Obendorf and C.M. Solbrig, "Distribution of Organophosphorus Pesticides on Cotton/Polyester Fabrics After Laundering as Determined by Electron Microscopy," International ASTM Symposium on the Performance of Protective Clothing, (American Society for Testing and Materials, Raleigh, NC, 1984); in ASTM STP 900, Performance of Protective Clothing, 1986.

9. C.N. Schlatter, "Permeation Resistance of Gloves After Repeated Cleaning and Exposure to Liquid Chemicals," presented at the American Industrial Hygiene Conference, Philadelphia, PA (1983).

10. J.L. Perkins, J.S. Johnson, P.M. Swearengen, C.R. Sackett, and S.C. Weaver, "Residual Spilled Solvents in Butyl Protective Clothing and Usefulness of Decontamination Procedures," *App. Ind. Hyg.* **2**:179 (1987).

11. C.E. Garland and A.M. Torrence, "Protective Clothing Materials: Chemical Contamination and Decontamination Concerns and Possible Solutions," Second International Symposium on the Performance of Protective Clothing (American Society for Testing and Materials, Tampa, FL, 1987); in ASTM STP Performance of Chemical Protective Clothing, Vol. 2 (1988).

12. J. Keaschall, J. Laughlin, and R.E. Gold, "Effectiveness of Laundering Procedures on Pesticide Removal Between Classes," International ASTM Symposium on the Performance of Protective Clothing, (American Society for Testing and Materials, Raleigh, NC, 1984); in ASTM STP 900, Performance of Protective Clothing, 1986.

13. D. Younger, Edmont Corp., private communication (1983).

14. C.N. Schlatter, "Effects of Dry Cleaning on the Performance Properties of Natural Rubber and Nitrile Rubber Gloves," presented at the American Industrial Hygiene Conference, Detroit, MI (1984).

15. K.C. Ashley, "Decontamination of Protective Clothing," International ASTM Symposium on the Performance of Protective Clothing (American Society for Testing and Materials, Raleigh, NC, 1984); in ASTM STP 900 Performance of Protective Clothing, 1986.

16. N.I. Sax, *Dangerous Properties of Industrial Materials* (Van Nostrand Reinhold, New York, 1984), 6th ed.

17. *Edmont Chemical Resistance Guide,* 3rd ed., technical marketing literature from Edmont, Coshocton, OH (1983).

18. J.R. Kominsky and J.M. Melius, "Fire Fighter's Exposures During PCB Electrical Equipment Fires and Failures," presented at the American Industrial Hygiene Conference, Dallas, TX (1986).

Chapter 9

Development of a Chemical Protective Clothing Program

S.Z. Mansdorf

Introduction

Use of chemical protective clothing (CPC) has grown tremendously in the last decade. This increase is due in part to: (1) regulations and standards requiring a greater reliance on the use of personal protective equipment; (2) the apparent cost/benefit of using personal protective equipment instead of more elaborate engineering controls; and (3) the need for totally protective suits for use in asbestos removal, hazardous waste disposal, and certain governmental and industrial operations where personnel exposure is considered especially hazardous.

Historically, most protective clothing was considered to be generically "impermeable" or liquid-proof and thought to be an absolute safeguard for the worker. Restrictions on the use of protective clothing were generally based on the salesperson's recommendations or simple immersion tests published in promotional brochures by the manufacturers themselves. This approach has been shown to be wholly inadequate for evaluating the level of protection provided.[1] Development of standard test methods that measure degradation, penetration, and permeation resistance of chemical protective clothing has resulted in the manufacturers, users, and researchers publishing specific recommendations for the selection of chemical protective clothing keyed to the materials of construction.[2,3]

Both routine and emergency handling of chemicals could result in direct skin exposure to toxic chemicals. Examples of such situations include:

- Handling of liquid chemicals during manufacture.
- Spray painting.
- Maintenance and quality-control activities for chemical processes.
- Acid baths and other treatments in electronics manufacture.
- Plating processes.
- Application of pesticide and agriculture chemicals.
- Emergency chemical response.
- Equipment leaks or failures.

The Occupational Safety and Health Administration (OSHA) requires that "... all personal protective equipment shall be of a safe design and construction for the work to be performed."[4] However, the publication of data demonstrating the permeation or penetration of protective clothing and gloves that have been generally believed to be "impermeable" requires a more comprehensive approach to the selection and use of CPC than is implied by the OSHA requirements.

Chemical protective clothing is intended to provide protection for personnel against hazardous chemicals when other more-effective methods of protection, such as engineering controls, are either inappropriate or infeasible. The possible configurations of CPC use can range from simple latex gloves to totally encapsulating suits, depending on the potentially affected body zone and the level of protection desired. Some configurations, such as totally encapsulating suits, can provide significant overall protection.[5] However, it should be recognized that the CPC user's actual level of protection will depend on the adequacy and effectiveness of the management system instituted by the provider to assure the proper selection, use, and maintenance of the protective clothing.

This chapter describes a systematic approach to the development of a comprehensive CPC management program.

Requirements for a Chemical Protective Clothing Program

The relative complexity of an individual CPC program will be determined by the level of protection required, the frequency and extent of CPC use, the risk to the users if the CPC fails, and the available staff and financial resources. This chapter has been written to address the high-risk situation where a formalized and comprehensive program is required; however, most of the program aspects described here are equally applicable to situations that are less potentially threatening.

Key Program Elements

The following sections describe a systematic approach to the development of a comprehensive CPC-management program. A comprehensive program should include eight key elements. They start with assessing the need for chemical protective clothing by reviewing the nature of the hazards of the process or job. This should be followed by determining the protection level and performance required of the CPC. With this hazard assessment, a proper CPC selection can be made.

Once the CPC has been selected, laboratory and field validation of the level of protection provided by the clothing and proper decontamination procedures should be conducted. This would normally be followed by training the user and his or her immediate supervisors in the limitations, proper decontamination, and maintenance of the CPC used. Supervisors would be expected to conduct regular inspections of CPC use, maintenance, and repair to assure the proper protection of the worker. Finally, a management audit scheme should be established to assure the continuing effectiveness of the overall CPC program.

Assessing the Need

The first step in determining the need for chemical protective clothing is to characterize the nature and extent of the potential hazard(s) in the workplace. This requires an initial health-hazard assessment conducted by a person competent in health and safety evaluation of the particular process or job. Normally, such assessments require an on-site inspection of the job or process to identify the potential stressors, their physical state and properties, routes of exposure and body zones of potential chemical contact, type of contact reasonably expected (e.g., splash), maximum expected concentrations for airborne contaminants, physical demands of the job, and the level of hazard associated with routine, intermittent, or potential accident exposure to the chemical agent(s).

The second step of this hazard assessment is evaluating alternatives to the use of chemical protective clothing. Alternative control approaches might include the following:

• Substitution of a less-hazardous chemical.

• Redesign of the process to reduce or eliminate worker contact with the hazardous chemical(s).

• Engineering controls (e.g., local exhaust ventilation) to reduce process emissions.

• Robotics or other mechanical materials-handling techniques.

Certain high-risk tasks and special personnel assignments, such as hazardous waste site work, spill response and cleanup, chemical fire control, and other such situations where alternative control techniques are not feasible, generally require a high level of personal protection by the CPC. This is especially true when the type and extent of chemical hazards are not known; in these situations, using a worst-case scenario may be required for the selection of chemical protective clothing. The Environmental Protection Agency (EPA) has developed guidelines on selection of personal protective equipment for site entry at hazardous-waste sites.[6] However, this scheme generally requires a knowledge of possible air concentrations of potential contaminants, and in many cases such information is simply not available. For these types of situations, a comprehensive approach to the development of a CPC program may be used to assure the proper protection of personnel at risk.

Development of Performance Criteria

Performance criteria for the selection of CPC should be based on an evaluation of the physical-resistance requirements of the protective clothing for the job, in addition to the level of chemical protection required. For example, will the garment or accessories be required to be flame retardant or thermally protective? Will the garment or accessories need to be especially abrasion- or tear-resistant? Job characteristics such as these should be profiled to assure that all factors are taken into consideration during the selection process. Chemical-resistance requirements should be based on the potential adverse effects of skin contact with the hazardous chemical. Acute effects may be simplified into five general classes as follows:

1. Minimal irritation or other non-permanent effect.
2. Moderate irritation or other non-permanent effect.
3. Severe irritation or other non-permanent effect.
4. Severe toxicity, burns, or other permanent effect.
5. Immediately dangerous to life.

Chemical protective clothing selected for a Class-1 hazard, such as skin contact with isopropyl alcohol, could simply be latex gloves, while Class-3 and -4 hazards, such as chromic acid exposure, require more complete protection from potential penetration, permeation, and degradation of the CPC. Note that chemical carcinogen hazards are not included in this rating scheme because their effects are chronic rather than acute; however, they would probably fall within the range of Classes 2–4, depending on the relative potency of the carcinogen.

Selection of CPC

The assessment of need, development of performance criteria, and hazard assessment will provide the basis for the selection of the level and type of protective clothing needed for a specific task or job. This information should be retained in writing in case the need to review the basis of the selection process arises. This could occur with the need for peer review of the selection process, failure of the selected items resulting in an accident or incident investigation, litigation based on an improper selection, or other occurrences requiring a formal review of the selection process.

The document on which this data is recorded should also include other information related to the selection and use conditions of the CPC. Figure 9-1 is a form containing all of the basic categories of information, which could be recorded and retained as part of the program files; Figure 9-2 shows an example of a completed form. This relatively simple form requires the

CHEMICAL PROTECTIVE CLOTHING SELECTION WORKSHEET

<table>
<tr><td colspan="2">Job Classification or Task:</td><td colspan="3">Process or Task Summary:</td></tr>
<tr><td colspan="4">Potential or Actual Hazards
Chemical:

Physical:</td><td>Contact Period:</td></tr>
<tr><td colspan="2">Type of Potential Contact:</td><td colspan="3">Body Zones of Potential Contact:</td></tr>
<tr><td colspan="5">Toxicology for Chemical Exposures:</td></tr>
<tr><td colspan="5">Potential Effects of SKIN Exposure:</td></tr>
<tr><td colspan="5">Permeation, Penetration, Degradation Data & Source:</td></tr>
<tr><td colspan="5">Recommended Base Materials of Construction for Each or All Component CPC:</td></tr>
<tr><td colspan="5">Specifications for Each CPC Component Required:</td></tr>
<tr><td colspan="5">Respiratory Protection:</td></tr>
<tr><td colspan="5">Type/Level of CPC Required:</td></tr>
<tr><td colspan="5">Training Required:</td></tr>
<tr><td colspan="5">Decontamination Procedures Required:</td></tr>
<tr><td>Worksheet Completed By:</td><td>Date:</td><td colspan="2">Checked By:</td><td>Date:</td></tr>
</table>

Figure 9-1. Chemical protective clothing selection worksheet.

CHEMICAL PROTECTIVE CLOTHING SELECTION WORKSHEET

Job Classification or Task:	Process or Task Summary:
Desmut Tank Cleaning-Maintenance Department	Drain tank in Weld Dept., Clean Sludge, Refill

Potential or Actual Hazards	Contact Period:
Chemical: Nitric Acid 30% v:v Hydrofluoric Acid 40% v:v (25% HF + Water) Water 30% v:v Physical: Heat stress and possible slip/trip hazard Also need head protection	Approximately 2 hours

Type of Potential Contact:	Body Zones of Potential Contact:
Potential splash while cleaning. Constant contact by boots. Otherwise intermittent.	Feet, Hands, Face (splash), Head (Fall), Body (splash)

Toxicology for Chemical Exposures: Causes severe burns to skin, eyes, and mucous membranes. Nitric acid vapor and oxides are insidious-symptoms delayed. Fumes are toxic-severe burns to respiratory tract. Immediate damage to eyes.

Potential Effects of SKIN Exposure: Immediate damage to skin-corrosive. Reaction products may also be corrosive. HF may cause delayed burns. HF burn treatment requires complexing of fluoride ion.

Permeation, Penetration, Degradation Data & Source: ACGIH Guide, 2nd Ed, Matrix A. Neoprene=R (HF) and Neoprene=r (Nitric). NIOSH database Neoprene for HF (25% solution) greater than 3 hrs breakthrough. Neoprene for nitric acid greater than 6 hrs breakthrough. Manufacturer MSDS recommends neoprene for nitric and PVC or neoprene for hydrofluoric. (Ajax Chemical Company- Madison, NJ)

Recommended Base Materials of Construction for Each or All Component CPC:

NEOPRENE: Gloves-double glove
Boots-to be sealed against legs
Head-hood
Suit-splash or full ensemble

Specifications for Each CPC Component Required:

See above-all parts of CPC to be Neoprene and sealed if components. Gloves and boots must be checked after use and deconed. Gloves to be thicker than 5 mils and lined.

Respiratory Protection:

Full facepiece airline respirator under positive pressure with escape air bottle

Type/Level of CPC Required:

Level B-Fully encapsulating suit with supplied air respirator

Training Required: Hazards of hydrofluoric acid and nitric acid. First aid procedures and decon. procedures. Rescue procedures. Requires safety person on-site

Decontamination Procedures Required: All equipment to be decontaminated after use by complexing agent for HF such as calcium chloride solution. Followed by water wash. All equipment to be tested for effective decon. before leaving area.

Worksheet Completed By:	Date:	Checked By:	Date:
S.Z. Mansdorf S.Z. Mansdorf	1/1/87	John Ensemble John Ensemble	1/4/87

Figure 9-2. Sample of completed worksheet.

person selecting the CPC to record information related to the job or task identifier; process or task description; potential or actual chemical and physical hazards; contact period; type of potential contact; body zones of potential contact; toxicology description for chemical exposures; potential effects of skin exposure; permeation, penetration, and degradation data and sources; recommended base materials of construction for each or all components; specifications for each CPC component; respiratory protection requirements; type and level of CPC required; worker training required; decontamination procedures; and signature and date blocks for the author and a reviewer.

The form shown in Figure 9-1 could be altered or modified to fit any special needs of the user or CPC program, and could easily be maintained in a computer database or simple notebook format.

Laboratory and Field Validation

Once the selection of candidate materials of construction for CPC has been completed, laboratory testing should be conducted to confirm the vendors' specifications. In many cases, this information may be extracted from data already published using standard methods, such as the American Society for Testing and Materials (ASTM) methods for permeation and penetration.[7,8] After completion of a literature review or laboratory studies, field evaluations should be conducted. This would include trials under actual field conditions of use. No published standard methods to determine the potential permeation or penetration of chemicals against CPC on the job site presently exist. However, Berardinelli and others have presented a number of proposed field testing techniques.[9]

Decontamination Procedures

Establishing effective decontamination procedures is important for both single-use and reusable chemical protective clothing. Contaminated single-use protective clothing may result in the transfer of that contamination to the wearer's clothing during doffing of the CPC. In this situation, single-use items should be decontaminated before the worker disrobes. Reusable CPC items will normally require some form of decontamination, and they should therefore be routinely checked for residual contamination on both the inside and outside surfaces before reuse.

Decontamination techniques vary depending on the reactivity and solubility of the contaminating chemical agent. Water-soluble chemicals may be removed with a detergent-and-water wash, while some chemicals may require complexing or other methods of inactivation before removal. Proper decontamination procedures should be predetermined and conveyed to the CPC wearer before his or her entry into the work area. See Chapter 8 for a more detailed discussion of CPC decontamination.

Training Program

All workers using chemical protective clothing should be trained in the use and limitations of their assigned equipment. This training should include, as a minimum, the following information:

- The nature, extent, and health effects of chemical hazards posed by the job.
- The proper use and limitations of the CPC assigned.
- Decontamination procedures, as appropriate.
- Inspection, maintenance, and repair procedures for the CPC.
- First-aid and emergency procedures.

Training should be conducted before actual use of the chemical protective clothing, and should allow actual "hands-on" experience. The trainer should be familiar with the specific CPC assigned and should also be able to answer questions. In order to measure the effectiveness of the training, a mechanism such as written pre-training and post-training tests should be included in the training plan. A roster of participants and an outline of the information presented should be maintained by the CPC provider. Retraining of CPC users should be conducted if any process or situation changes affect the use of CPC, and at yearly intervals for refresher training.

Inspection, Maintenance, and Repair

A routine inspection program should be established, to be conducted on three levels. The first level should be routine worker inspection of the CPC before and after each use. The second level should include scheduled supervisory inspections of the chemical protective clothing. The third level should include an audit of these functions by the management person designated responsible for the CPC program. Maintenance should be conducted in a similar manner. Workers should be responsible for turning in defective or worn equipment to their supervisors. Supervisors should also inspect the protective clothing for any additional defects or wear before turning each item in for repair or replacement.

All major repair work should be done by a fully competent person who is able to certify that the condition of the repaired gear is equal to or better than the original manufacturer's specifications.

Auditing the Program

Management should establish a mechanism to assure that all elements of the CPC program are carried out effectively. Typically, this consists of an audit procedure where competent personnel independent of the daily operation of the CPC program review documentation and standard procedures on a yearly basis, providing recommendations for improvement. This peer review is an integral part of maintaining an effective program.

Summary

Chemical protective clothing is intended to provide protection for personnel against hazardous chemicals when other more-effective methods of protection, such as engineering controls, are either inappropriate or infeasible. Protective clothing can range from simple latex gloves to totally encapsulating suits.

The actual protection offered by chemical protective clothing will depend on the adequacy and effectiveness of the established management system to assure the proper selection, use, and maintenance of the CPC. A comprehensive program should include:

- Assessing the need for CPC.
- Determining the protection level and performance required of the CPC.
- Proper selection of CPC based on the hazard assessment.
- Lab and field validation of the level of protection provided by the CPC.
- Establishment of effective decontamination procedures.
- Training of personnel in the proper use and limitations of the selected CPC.
- The routine inspection, maintenance, and repair of CPC.
- Establishing a management audit scheme to assure the effectiveness of the program.

References

1. S.Z. Mansdorf, "Risk Assessment of Chemical Exposure Hazards in the Use of Chemical Protective Clothing—An Overview," *Performance of Protective Clothing,* ASTM STP 900, R.L. Baker and G.C. Coletta, Eds., American Society for Testing and Materials, Philadelphia, PA (1986), pp. 207–213.

2. A.D. Schwope, P.P. Costas, J.O. Jackson, and D.J. Weitzman, *Guidelines for the Selection of Chemical Protective Clothing,* 3rd ed. (American Conference of Governmental Industrial Hygienists, Cincinnati, OH, 1987).

3. S.Z. Mansdorf, "Chemically Resistant Glove Use Helps Prevent Skin Contamination," *Occupational Health & Safety,* **56**:79–83 (1987).

4. Occupational Safety and Health Administration, *General Industry Standards,* 29 CFR 2910.132(c), Occupational Safety and Health Act of 1970 (84 Stat. 1593), Government Printing Office, Washington, DC (revised 1978).

5. A.D. Schwope and R.E. Hoyle, "Tame Hazardous Waste Hazards with Personal Protective Equipment," *Hazardous Materials and Waste Management* **3**:14–22 (1985).

6. Environmental Protection Agency, "Interim Standard Operating Safety Procedures," US Environmental Protection Agency, Office of Emergency and Remedial Response, Hazardous Response Division, Washington, DC (1982).

7. American Society for Testing and Materials, "Standard Test Method for Resistance of Protective Clothing Materials to Permeation by Hazardous Liquid Chemicals," ASTM Method F739, American Society for Testing and Materials, Philadelphia, PA (1987).

8. American Society for Testing and Materials, "Standard Test Method for Resistance of Protective Clothing Materials to Penetration by Liquids," ASTM Method F903, American Society for Testing and Materials, Philadelphia, PA (1985).

9. S.P. Berardinelli and M.M. Roder, "Chemical Protective Clothing Field Evaluation Methods," *Performance of Protective Clothing,* ASTM STP 900, R.L. Baker and G.C. Coletta, Eds., American Society for Testing and Materials, Philadelphia, PA (1986), pp. 250–260.

Index

O

P

R

S

T

V

W

Z